Lionel Ernest Adams

The Collector's Manual of British Land and Freshwater Shells

Containing figures and descriptions of every species, an account of their habits and

localities, hints on preserving and arranging, etc. The names and descriptions of all

the varieties

Lionel Ernest Adams

The Collector's Manual of British Land and Freshwater Shells
Containing figures and descriptions of every species, an account of their habits and localities,
hints on preserving and arranging, etc. The names and descriptions of all the varieties

ISBN/EAN: 9783337218287

Printed in Europe, USA, Canada, Australia, Japan

Cover: Foto ©Andreas Hilbeck / pixelio.de

More available books at **www.hansebooks.com**

THE
COLLECTOR'S MANUAL OF BRITISH
LAND AND FRESHWATER
SHELLS.

THE

COLLECTOR'S MANUAL

OF

BRITISH LAND AND FRESHWATER

SHELLS.

CONTAINING FIGURES AND DESCRIPTIONS OF EVERY SPECIES, AN
ACCOUNT OF THEIR HABITS AND LOCALITIES, HINTS ON
PRESERVING AND ARRANGING, ETC.; THE NAMES
AND DESCRIPTIONS OF ALL THE VARIETIES
AND SYNOPTICAL TABLES SHOWING
THE DIFFERENCES OF SPECIES
HARD TO IDENTIFY.

BY

LIONEL ERNEST ADAMS, B.A.,

AUTHOR OF " A CONCISE SYSTEM OF ENGLISH PARSING."

ILLUSTRATED BY GERALD W. ADAMS AND THE AUTHOR.

LONDON:
GEORGE BELL AND SONS, YORK STREET,
COVENT GARDEN.
1884.

INTRODUCTION.

THE object of this little "manual" is to enable a novice to collect, identify, and arrange systematically the various shells—both land and freshwater—which abound in every part of these islands.

Remembering the difficulties that I at first encountered in identifying the various species, I have paid special attention to those which are likely to present themselves to other beginners.

To this end I have framed synoptical tables of the differences between those species of *Zonites* and *Vertigo* which are most closely allied.

I have also shown in a tabular form the relations the different classes, orders, and genera bear to one another.

I have taken as a model the "British Conchology" of Dr. Gwyn Jeffreys—our highest authority on the subject. I am also indebted to Forbes and Hanley's work, Mr. Rimmer's excellent book, to Mr. T. Rogers of Manchester, and to several others for many useful facts and hints.

With the exception of the *Pupæ* and *Vertigos*, which I have myself enlarged from actual specimens, the plates

B

have been drawn by my brother Gerald W. Adams, a collector like myself. Most of the drawings have been taken from shells in my own cabinet; here and there, however, a friend has been generous enough to lend me some fine or rare specimen to copy.

Where there is no "size-line" the figures may be taken as life size, except in the case of the three largest bivalves, which for convenience sake I have figured smaller than the average size of adult specimens. In these cases I have stated their dimensions beside the figures, and also in the descriptions.

I have translated and accentuated the specific names of all the species, and have appended a glossary of all the technical terms which I have been compelled to make use of.

It should be borne in mind that plates and descriptions are at best but a second-rate substitute for a direct examination of the objects themselves, and that far more may be done towards the identification of difficult species by careful comparison with a good collection of well-authenticated specimens, than by working at plates and* descriptions alone; just in the same way that more correct and useful knowledge of anatomy may be gained and fixed in the memory by a couple of hours' dissecting than in a week of poring over the best diagrams.

Most of our large towns, and many of the smaller ones, boast of some sort of museum where British shells have a

place, and should any stranger wish for further information than can be obtained through the glass cases, the
curator will always be ready to give courteous attention to
his enquiries.

Though the entire list of British land and freshwater
species is limited to 130, there are numerous varieties of
most of these; so that while a fair collection is within the
reach of most people—often within the range of a single
county—a *perfect* collection takes a long time to accu-
mulate.

The pleasures of collecting *anything* are too patent to
need to be dwelt on here, but the pleasures of collecting
objects of natural history in any branch have additional
charms—the charms of the country.

It should not be supposed that the summer is the only
time when it is possible to collect. Throughout the winter
much may be done, except, of course, when a frost puts an
end to turning over stones, and drives all water shells into
the mud.

Doubtless every describer of shells has been puzzled to
find simple geometrical terms suitable to their forms, which
are often complex. I have followed the authorities in
using such terms as *ovate, subcylindrical,* &c., though they
are vague, and do not express to what degree the object is
oval or cylindrical. Dr. Jeffreys calls *L. stagnalis* "elongated," which it certainly is; but what *shape* is "elon-
ated"? This difficulty, however, affects the describer

more than the collector, who can glance at a plate where the shell is figured. Some ludicrous results of attempting to realize form from description may be seen in the drawings of the old naturalist Gesner, who depicts elephants, whales, besides other beings more fearful and wonderful still, from the descriptions of people who had seen them, or professed to have done so.

Shells recently admitted into the British List.

Within the last few years several shells have made their appearance in these islands.

Some of these, whose introduction from foreign parts is fairly established, have been admitted to rank with indigenous species. These are *P. dilatatus*, which came in American cotton bales, and *T. Maugei*, which is shown to have been introduced with continental plants sent to Bristol. Both of these species have made themselves at home, and are spreading.

Others, known hitherto only as foreign, and whose method of introduction is obscure, have been admitted by some as British species. Among these are *Clausilia parvula* and *Helix villosa*.

Zonites glaber, Vertigo Lilljeborgii, and *V. Moulinsiana,* though they had escaped notice till quite recently, can hardly have been introduced by the ordinary methods, and may fairly claim to be indigenous.

A few long-established British species, whose introduc-

tion is suspected to be comparatively recent, are *S. ovale*,
D. polymorpha, *H. Pisana*, and *H. obvoluta*.

It was thought that *H. pomatia* was introduced by the
Romans as an article of food, but this notion is now discarded.

Ways and Means of Collecting.

I have often been asked by would-be collectors such
questions as "How do you set about collecting?"
"What implements are necessary?" and most frequently,
perhaps, "Where do you look for shells? I know the
'garden snail,' and a yellow one with bands, but I never
see all these you have in your cabinet."

To those desirous of this and similar information the
following hints may prove serviceable :—

When going out for a ramble after shells very few pre-
parations are necessary. For pond work a scoop is essen-
tial, and better than a net. The most handy scoop is one
of very fine zinc gauze, with a sharp zinc rim, to which is
attached an open ferrule, into which a walking-stick may
be inserted. A store of tin boxes of various sizes is re-
quired, and in selecting these it should be remembered that
those which open with a spring are more handy than those
the lids of which have to be removed every time a shell is
dropped in. Small nib-boxes and match-boxes are very
good for small shells, and should be padded with a little
weed or moss to prevent the more delicate species being
broken by being rattled together in the pocket. *Tin* boxes

are much better than "chip" boxes or pill-boxes, as they
are less likely to get broken, and do not come to pieces
when wet. Hispid shells should be put into a box by
themselves, and only a few together, as their neighbours'
slime is apt to spoil their personal appearance. For the
minute species of freshwater shells a small wide-necked
bottle filled with water may be found useful to dip the
fingers in, and so wash off the shells, which often adhere
persistently, and require much time to dislodge in safety.
The shells will sink to the bottom, and the water may be
poured away. For the minute and delicate species of both
land and water, it is not a bad plan to keep one or two
small glass tubes in the waistcoat pocket. Along one side
of these a strip of gummed paper (stamp-paper is very
good) should be fastened to hold the glass together if
cracked by a fall. Such tubes are obtainable at a homœo-
pathic chemist's, at eightpence per dozen.

Finally, remember never to be without a receptacle of
some sort when out, even though not on a regular ex-
pedition. Should you happen to be thus unprovided, you
will be sure to regret it.

In the case of the *Zonites*, let the collector gather all he
can till he knows them well, and he will often find, when
he examines them at leisure, that he has entertained an
angel unawares in the shape of some good variety.

No ponds or ditches should be passed by without examination, however barren they may appear, and not only

should the weeds be examined, but the mud should be sifted with the scoop in search of bivalves.

By the water's edge the stalks and leaves of flags and sedges should be examined for the *Succiniæ*, which are amphibious.

On land, search all moist and shady spots, especially during and after rain, under logs, stones, among dead leaves and decaying vegetation, among nettles and healthy vegetation, on the bark of trees, and at their roots among the moss, on old stone walls, and in damp cellars. The rejectamenta of rivers, too, yield a fruitful harvest.

A good plan for dealing with dead leaves and moss is to take a quantity home, spread it out to dry, and search the siftings. This saves much time, and often yields a good supply of *Zonites*, minute *Helices*, *Pupæ*, &c. But this should be done with judgment—as with the moss the eggs also come away, and the habitat is destroyed.

It often happens that we come across a good shell in a likely place, and wherever we find a single individual we may be pretty certain that some of his immediate relatives are not very far distant. A large, flat stone, log, or piece of matting laid over the place will frequently be found on examination, after a day or two, to have the desired object adhering to its under side. I have frequently set "traps" of this nature, which I visited periodically, and which have been very productive.

It may be remembered that sandy or peaty soils yield

little or nothing (though a great exception must be made in favour of sand-hills by the sea coast) ; nor do shells live in pine-woods—the resin, perhaps, being distasteful to them; nor among bracken. Calcareous districts, on the other hand, are especially rich fields for search.

Preparing Shells for the Collection.

The " booty " should be cleaned as soon as possible after being captured. Plunge the shells into boiling water, and extract the animal with a pin. My most serviceable extractors for small species are fine needles stuck head first into the wood of common matches. The points of the needles are bent into curves of different shapes and sizes to reach the interiors of shells into which the animals sometimes shrink or remain broken. Care should be taken to wash the mouths of small shells with a paint-brush. The larger water shells are often improved by a gentle application of soap and hot water with a moderately soft tooth-brush; but it is a fatal mistake to use *acid* in any form.

It will be found, especially at first, that the animals of many shells break inside, which spoils the appearance of the transparent *Zonites,* &c. This cannot always be helped, but may to some extent be avoided if the animals are drawn out, before they cool, very slowly and steadily. They should not be boiled too hard, but only plunged for a few seconds into boiling water.

The animals of very minute species need not be removed —the shells may be simply dried.

A penknife carefully inserted will separate the animal from the shells of the bivalves, which should be instantly tied up or screwed up in a piece of tissue paper till dry. Should this not be done, the ligament will harden with the valves open, and the shell cannot then be closed without snapping the ligament.

A set of slugs preserved in spirits forms an interesting feature in a collection; and some guidance is necessary here.

The animals must not be plunged alive into spirits and straightway sealed up. If this is done, they will exude a thick coat of mucus, which surrounds them like a cocoon, and, moreover, they will shrink up to a very small uninteresting mummy.

They should be drowned in cold water, and when dead should be cleaned of the inevitable mucous coat with a paint-brush. They should then be put into methylated spirit and water (in proportion of 1 spirit to 3 water). After three or four days they should be again wiped and transferred to a mixture of equal spirit and water, and finally, after another interval of the same length, to a mixture of 3 spirit to 1 water.

By this means their tendency to shrink is minimized, and they are more apt to retain their markings, which pure spirit is apt to obliterate.

The various slugs in neat glass tubes arranged in trays

on cotton wool, carefully labelled, have a very attractive
appearance in a collection.

The shells of most slugs are covered by the mantle, under
which the blade of a penknife or scalpel should be inserted
to effect their removal. Many prefer to kill the animals
first in boiling water, and this method recommends itself
both on the score of humanity and convenience. The shells
thus extracted may be gummed on to slips of black card.

Labelling and Registering.

As soon as possible the shells should be labelled, even if
they are duplicates, to be stored away for exchange. Both
the *name* and *locality* should be most carefully recorded.

The collector will do well to keep a register. There are
various methods of arranging this, but the following has
commended itself to me as the most practical:—Get a
good-sized note-book, keep a separate page for each species
with the name written at the top, and the localities in a
marginal column. I give a specimen :—

Cochlicopa tridens.

Matlock.	A few specimens found among moss at the foot of limestone rocks at Matlock Bridge. Sept., 1880.
Coggeshall.	Sparsely among dead leaves. 1881.
Evesham.	Specimens received, coming from Evesham. 1881.
Marple (Cheshire).	A single specimen in moss on a wall. Aug., 1882.

&c., &c.

There should be some spare pages at the end of the
register for occasional notes. For instance, after a district
has been well worked, a list of all species found in it

should be made, and the extent of the district explored should be accurately stated.

The Arrangement of a Collection.

Much of the pleasure derived from a collection consists in its arrangement. A cabinet with drawers is undoubtedly the best receptacle, but a series of flat boxes will answer the purpose very well.

Various methods of setting out the specimens are adopted. Some prefer slabs of wood or glass covered with stone-coloured paper, on which the shells are fastened with gum. This is doubtless an excellent plan for showing off the shells, especially in a museum or a perfect collection where the shells are not supposed to be touched; but in the case of a collection which is constantly receiving better specimens to be substituted for those already stuck down, this method has an obvious drawback. My own experience, in the case of all but very small shells, is in favour of card-board trays, lined with cotton wool, on which the shells are easily arranged and are well shown out. The tray method has the advantage of allowing the specimens to be easily changed, and, if necessary, taken up and handled. These trays may be obtained from many London naturalists.

Pink cotton wool, though in favour with some, has not a pleasing effect.

Minute shells may be sometimes kept loose in the small glass tubes before mentioned, or in small glass-topped boxes, which can be handled without endangering the contents. Sometimes a set of minute shells may be fastened to a strip of card, which strip may be put into a tube or glass-topped box. Some delicate white shells, such as *Z. crystallinus*, *H. pulchella*, *A. acicula*, &c., are better shown on black card. In all cases where shells are stuck down gum tragacanth should always be used, as it does not glaze when dry like gum arabic. In every case a label on which is recorded the *name* and *locality* should be affixed.

Varieties.—Most species have one or more well-marked varieties of form and colour. Though it is of great importance to secure specimens that vary from the types, such specimens should never be labelled or registered with any varietal name without some good authority.

A *caution* may be useful to the collector not to admit into his collection any shell whatever without the most conclusive evidence of its being British. I have repeatedly been presented with foreign specimens by persons of the highest integrity with the assurance that they came from such and such a place in the British Isles. On one occasion a lady gave me some West Indian sea-shells, which she positively averred she had picked up in Jersey. I am sure the majority of collectors have undergone similar experiences.

In order that a collection may be of any value, it should be, like Cæsar's wife, " above suspicion."

Plate IX

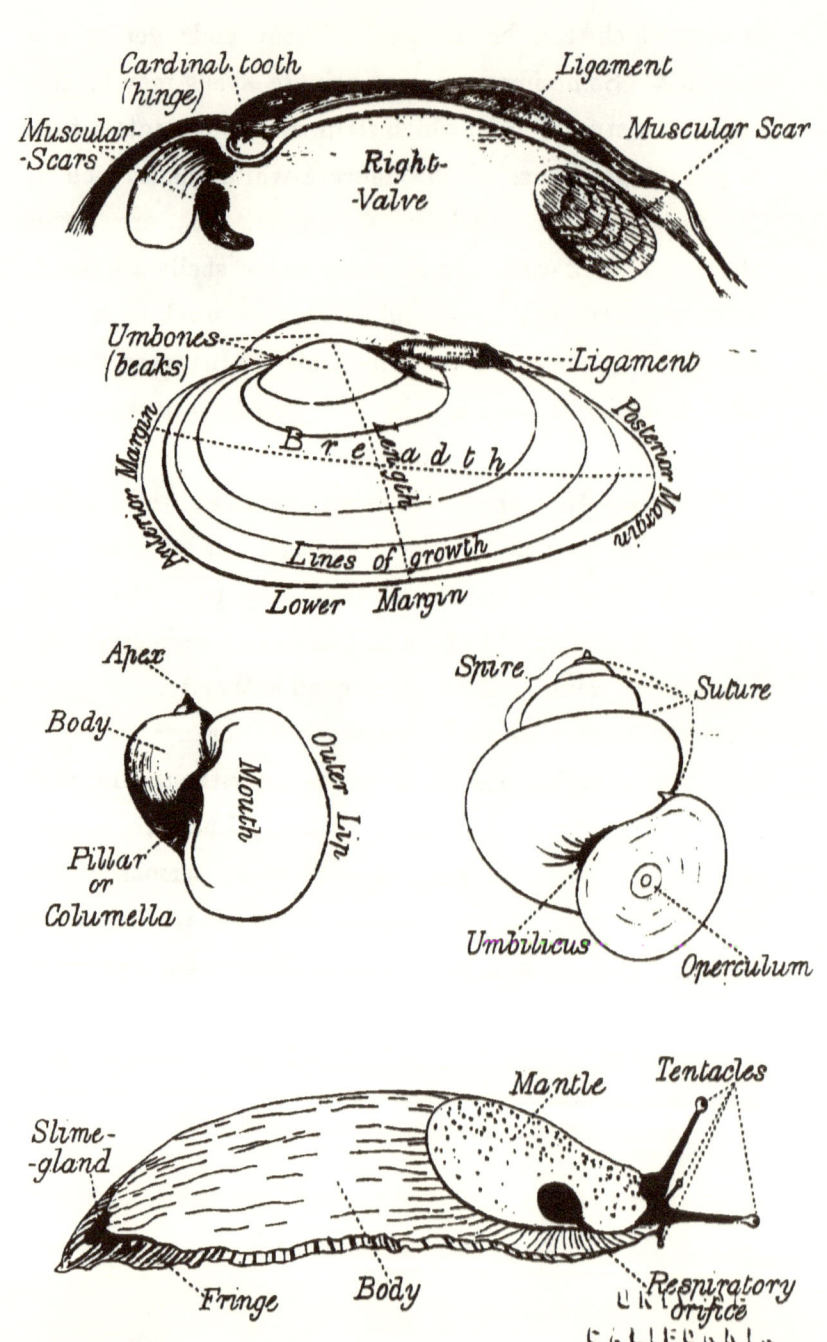

Cardinal tooth (hinge)

Ligament

Muscular-Scars

Muscular Scar

Right-Valve

Umbones (beaks)

Ligament

Anterior Margin

Breadth

Length

Posterior Margin

Lines of growth

Lower Margin

Apex

Body

Mouth

Outer Lip

Pillar or Columella

Spire

Suture

Umbilicus

Operculum

Mantle

Tentacles

Slime-gland

Respiratory orifice

Fringe

Body

G.W. Adams del.

G. Jarman sc.

PART I

AQUATIC.

CLASS I.—CONCHIFERA, OR BIVALVES.

Order.—LAMELLIBRANCHIATA (*with leaflike gills*).

Family I.—SPHÆRIDÆ.

Genus I.—SPHÆRIUM.

1. S. CÓRNEUM (*horn-coloured*).

Shape somewhat globular; thin, glossy, opaque, yellowish horn-colour, with bands of a paler tint indicating the lines of growth, striated finely and evenly in the same direction, also faintly striated from the beaks to the margin; *ligament* short; *hinge* furnished with a double cardinal tooth in each valve—two lateral teeth in the right, and four in the left valve; muscular *scars* faint.

This common shell is found at the roots of weeds, and in the mud in ponds, ditches, canals, and rivers in every part of the British Isles.

It varies considerably as follows :—

Var. I. *flavescens.* Smaller, more globular, and paler.

Var. II. *nucleus.* Smaller, nearly spherical.

Var. III. *Scaldiana.* Ovate, and paler than type.

Var. IV. *Pisidioides.* Somewhat triangular.

2. S. RIVÍCOLA (*inhabitant of streams*).

Oval, solid, opaque, glossy, greenish or yellow when young; *striæ* evenly concentric; *ligament* short, but conspicuous; *scars* distinct; *teeth* as in last species, but larger.

It inhabits canals and slow rivers in various parts of England, and near Dublin.

This is the largest species in the genus. It is also much flatter than *S. corneum,* especially when young, more solid and more strongly striated.

3. S. OVÁLE (*oval*).

Oblong, thin, compressed, semi-transparent, dull, pale grey; *striæ* faint, concentric; *beaks* central; *ligament* long; *hinge* straight on posterior, incurved on anterior side; *teeth* small; *scars* distinct.

This is a local species, being found in ponds and canals in Lancashire, Yorkshire, and near Birmingham. Its oblong shape, thin delicate appearance, pale colour, and straight hinge, prevent its being mistaken for any others of this genus.

4. S. LACÚSTRE (*inhabiting lakes*).

Round, more compressed than *S. corneum,* thin, glossy, semi-transparent, regularly striate concentrically; *beaks* small, but prominent; *ligament* narrow; *hinge* strong; *teeth* small; *scars* faint.

In similar habitats to *S. corneum*, but more local though widely distributed; Glasgow is the only Scotch locality. Its prominent beaks form its characteristic distinction from the other three species. It is, besides, much more compressed than *S. corneum*, and its "shoulders" are sharp, like those of *S. ovale*, though less pronounced. The finest specimens I possess were taken from the surface of the mud of a small pond which had been dried up for two months. All the specimens were alive.

Var. I. *Brochoniana.* Much larger and flatter.

Var. II. *rotunda.* Rounder and flatter.

Var. III. *Ryckholtii.* Small, triangular, and globular; *beaks* prominent.

Genus II.—PISÍDIUM.

These little shells are usually a great trouble to the collector on account of their variability and similarity, and conchologists have differed greatly as to the number of species. The thanks of all collectors, therefore, are due to Dr. Jeffreys, who has reduced the number to five fairly well-marked species, distinguishing them as follows :—

"A. Triangular (*P. amnicum, P. fontinale*).

B. Oval (*P. pusillum*).

C. Round (*P. nitidum*).

D. Oblong (*P. roseum*)."

The *Pisidia* differ from the *Sphæria* in having a single siphon, while the *Sphæria* have two.

1. P. ÁMNICUM (*inhabiting rivers*).

Triangular, tumid, solid, glossy, quite opaque when adult, having deep concentric ridges, greyish horn-colour; *teeth* (as are those of all the *Pisidia*) the same as those of the last genus.

P. amnicum is found in ponds, canals, rivers and lakes throughout the country. There is no mistaking this species when adult on account of its much greater size than any of the others; its triangular shape distinguishes it from the members of the genus *Sphærium.* Young specimens need not be mistaken for *P. fontinale,* as the latter shell is not so deeply groved, and far more ventricose.

2. P. FONTINÁLE (*inhabiting springs*).

Triangular; so tumid that it has a cuboid appearance; thin, greyish horn-colour; more transparent than *P. amnicum; beaks* prominent; *scars* deep.

Found in sluggish streams, canals, ponds, and ditches throughout the country.

Var. I. *Henslowana.* The valves have a plate-like appendage near the beaks.

Var. II. *pulchella.* Strongly grooved, glossy.

Var. III. *pallida.* More tumid, paler, irregularly striate.

Var. IV. *cinerea.* Larger and flatter.

3. P. PUSÍLLUM (*small*).

Oval, thin, moderately glossy, finely striate concentrically, yellowish white; *beaks* nearly central; *ligament* inconspicuous.

Common in ditches and swamps. Its oval form, as well as its blunter and more central beaks, distinguish it from the last species.

Var. I. *obtusalis.* Smaller and more tumid.

Var. II. *grandis.* Much larger (near Manchester).

4. P. NÍTIDUM (*shining*).

Roundish, tumid above, compressed below, very glossy; *scars* distinct.

It is found in ponds throughout the country. Its outline is rounder and its epidermis more glossy than the rest of the *Pisidia.* Its siphon is funnel-shaped.

Var. I. *splendens.* Lemon-coloured.

Var. II. *globosa.* Sphæroidal in form.

5. P. RÓSEUM (*rose-tinted*).

Somewhat oblong, tumid, glossy, regularly striate concentrically; *ligament* almost invisible.

Its habitats are similar to those of the last species. Its oblong shape, the situation of the beaks, which are on one side like those of *P. fontinale*, and the straightness of its lower margin, serve to mark it off from *P. nitidum*, which it somewhat resembles.

Family II.—UNIONIDÆ.

Genus I.—ÚNIO.

1. U. TÚMIDUS (*swollen*).

Ovate, very solid, dark brown; *epidermis* smooth.

Found in rivers, ponds, and canals in England and Wales commonly.

Var. I. *radiata*.　Thinner, radiated with yellow.

Var. II. *ovalis*.　Wedge-shaped, dark coloured.

2. U. PICTÓRUM (*Painters'*).

Oblong, solid, usually narrower than *U. tumidus*. The *ligament* is parallel with the lower margin, whereas the *ligament* of the last species, if produced, would meet the line of the lower margin, and form an angle. (See plate.) Its colour is usually a much brighter green, and it is more glossy than the last species. It is also narrower in proportion to its size, and less solid.

This shell is found in similar localities to *U. tumidus*, and often approaches it in appearance—so much so that there has been great doubt of its being a distinct species. Mr. Rogers informs me that specimens found in canals near Manchester are lined with a beautiful salmon-coloured nacre, due perhaps to the refuse at the bottom of the canal.

Var. I. *radiata.* Having rays of a greenish colour.

Var. II. *curvirostris.* Smaller, shorter, and flatter.

Var. III. *latior.* Broader and shorter, yellow-brown.

Var. IV. *compressa.* Very broad and flat.

3. U. MARGARÍTIFER (*pearl-bearing*).

Oblong, very solid, dull black; *beaks* always eroded, lower margin incurved; very compressed. The inside is pearl white, sometimes pinkish. Occasionally pearls are found inside—white, green, or brown. Length two inches, breadth five inches.

This interesting species is to be found in rivers in mountain districts in several parts of Great Britain, and also in Ireland and Man. The pearls for which this shell was once eagerly gathered in the Tay, the Irt, and the Conway, are small and worthless compared with those from the East.

Suetonius says that Cæsar was partly attracted to

Britain by the reports of pearls found there, and Pliny states that he covered a buckler with them, which he dedicated to Venus Genetrix.

Forbes and Hanley think that "Cæsar's buckler was more probably covered with the pearls from *Mytilus edulis*" (the common sea mussel). This, however, is not likely, as the pearls from this shell are exceedingly few and poor. Tacitus writes that they were of marine origin.

Pennant states that as many as sixteen pearls have been found in a single *Unio*, and he gives an account of pearls of value having been found in Donegal and in the Conway.

The ancient writers agree in disparaging the British pearls, justly considering those from the East finer in size and quality.

Tacitus mentions a theory current in his time that the dull reddish colour of our pearls was due to their being collected from cast-up shells instead of being gathered from living shells from the bottom of the sea; but he adds, with characteristic dry humour, that the fault probably lay in the pearls themselves, as otherwise his avaricious countrymen would have been sure to discover the best method of obtaining them.

Var. I. *sinuata.* Broader than the type, yellowish; lower margin incurved.

Var. II. *Roissyi.* Longer than the type, lower margin rounded outwards.

Genus III.—ANODÓNTA.

The members of this genus are almost toothless, as the name implies. They are ovoviviparous.

1. A. CŶGNEA ([*eaten by*] *swans*).

Oblong, swollen, thin, brown or green, well marked by lines of growth; *ligament* long and parallel to the lower margin; inside pearly, iridescent. Length two and a half inches, breadth five.

This is the largest of our freshwater mussels, attaining an average breadth of over five inches. It is found ·in great abundance in slow rivers, ponds, and canals as far north as Perth.

Var. I. *radiata.* Larger, rayed with yellow.

Var. II. *incrassata.* More swollen and solid.

Var. III. *Zellensis.* Broader, yellow-brown.

Var. IV. *pallida.* Light yellow, wedge-shaped.

Var. V. *rostrata.* Ovate, upper margin forming a crest.

2. A. ANATÍNA ([*eaten by*] *ducks*).

Oval, compressed, green or brown, glossy; *ligament* short, prominent, straight, forming an angle with the lower margin. (See plate.) *Scars* deeper than those of the last species. Length two inches, breadth three and a half inches.

Found in the same localities as *A. cygnea*. This species may be distinguished from the last by its smaller size and the angle its hinge line makes with the lower margin.

Var. I. *radiata*. Radiated with green and yellow.

Var. II. *ventricosa*. Larger, more solid.

Var. III. *complanata*. Oval, compressed, beaks close to anterior margin.

Family III.—DREISSENIDÆ.

Genus.—DREISSÉNA.

1. D. POLYMÓRPHA (*many-shaped*).

Shell triangular, sharply keeled, lower margin curved inwards, solid; *ligament* long and narrow. Animal furnished with a *byssus*.

There is no mistaking this shell. Its triangular carinated shape is sufficient to identify it at once.

It inhabits rivers, lakes, and canals in several parts of England. I have found it attached to iron pipes in a reservoir near Manchester. Fine specimens are found in the Thames.

CLASS II.—GASTEROPODA (*belly forming the foot*) OR UNIVALVES.

Order I.—PECTINIBRANCHIATA.

Family I.—NERITIDÆ.

Genus.—NERITÍNA.

N. FLUVIÁTILIS (*inhabiting rivers*).

Ovate, solid, glossy, chequered brown and white and purple; *whorls* three; *spire* short; *operculum* semi-lunar.

This prettily marked shell is found in many parts of Great Britain attached to stones in running water and canals. It is the most solid of the freshwater shells, and reminds us of its marine cousins, the *Littorinæ*, so common on seaweed.

Family II.—PALUDINIDÆ (*living in marshes*).

Genus I.—PALUDÍNA.

1. P. CONTÉCTA (*covered by operculum*).

Conical, diaphanous, fairly glossy, light green, with darker bands of the same colour; *whorls* seven, very convex; *suture* very deep; *mouth* almost circular, but angulated above; *umbilicus* small, but deep and distinct; *operculum* thin.

It inhabits slow rivers, canals, and ponds as far north as Lancashire and Yorkshire. It is often found coated with confervoid, which can easily be removed by hot water and soap applied with a soft nail-brush. Both this and the following species are viviparous.

2. P. VIVÍPARA (*producing young alive*).

Conical, but more oval than last species. Not so glossy as *P. contecta;* lighter in colour, and more solid. *Whorls* six and a half; *suture* not so deep as *P. contecta; umbilicus* none ; *operculum* moderately thick.

With the exception of one locality in Scotland (near Moray Firth) this species is found within the same limits as *P. contecta.* The two species may be distinguished by the following characteristics :—

P. CONTECTA.	P. VIVIPARA.
Conical.	More oval.
Rather glossy.	Not so glossy.
Dark green.	Light greenish yellow.
Suture very deep.	Not so deep.
Umbilicus distinct.	Umbilicus none.
Texture thin.	More solid.
Apex sharp.	Apex blunt.

Var. I. *unicolor.* Without bands.

Var. II. *atro-purpurea.* With dark purple markings.

Genus II.—BYTHÍNIA.

1. B. TENTACULÁTA (*with tentacles*).

Somewhat conical, rather solid, glossy, semi-transparent, horn-coloured; *whorls* six, convex; *suture* moderately deep; *mouth* oval, angulated above; *umbilicus* hardly perceptible; *operculum* thick.

Found in slow rivers, canals, ponds, &c., distributed throughout every part of the British Isles.

Var. I. *ventricosa.* Shell white, tumid.

Var. II. *decollata.* Upper whorls wanting.

Var. III. *excavata.* More tumid, suture deeper.

2. B. LEACHÍI (*after Dr. Leach*).

Conical, rather thin, moderately glossy, semi-transparent, horn-colour; *whorls* five, very convex; *mouth* almost circular, with slight angulation above; *umbilicus* small, but distinct; *operculum* as in last species.

This species is far more local and less abundant than the last, though found in similar situations. It occurs in Berks, Hants, &c. As it is only about half the size of *B. tentaculata*, adult specimens cannot be mistaken; but in case it should be mistaken for the young of that species the following distinctions may be useful:—

B. TENTACULATA.	B. LEACHII.
	Whorls much more convex.
	Suture much deeper.
Umbilicus practically none.	Umbilicus distinct.
Mouth obliquely oval and sharply angulated.	Mouth almost circular.

Genus III.—HYDRÓBIA.

1. H. SÍMILIS (*similar to another species*).

Sub-conical, thin, semi-transparent, yellowish horn-colour; *whorls* five to six, rounded; *suture* deep; *mouth* and *operculum* oval; *umbilicus* small.

This species is only found between Greenwich and Woolwich, in ditches which are occasionally flooded by the tide.

2. H. VENTRÓSA (*swollen*).

Conical, tapering, thin, glossy, semi-transparent; *whorls* six to seven, rounded; *spire* pointed; *mouth* oval.

Abundant in brackish water in estuaries in England and Wales.

Var. I. *minor*. Much smaller, spire shorter.

Var. II. *decollata*. Truncate, eroded.

Var. III. *ovata.* Four whorls, shorter, more tumid.

Var. IV. *elongata.* Sometimes eight whorls, longer.

Var. V. *pellucida.* White, transparent.

Family III.—VALVATIDÆ.

Genus.—VALVÁTA.

1. V. PISCINÁLIS (*inhabiting fish ponds*).

Somewhat globular, rather solid, nearly opaque, brownish yellow; *whorls* six, very convex; *spire* blunt; *suture* very deep; *mouth* circular; *umbilicus* deep.

Found in lakes, canals, and rivers throughout the British Isles.

Var. I. *depressa.* Flatter, umbilicus larger.

Var. II. *subcylindrica.* Spire more raised.

Var. III. *acuminata.* More raised and pointed.

2. V. CRISTÁTA (*crested, i.e. with branchial plume*).

In shape a circular coil, glossy, light horn-colour, rather solid; *whorls* five; *mouth* circular; *umbilicus* very large and open.

This species frequents canals, lakes, ponds, &c. It may generally be found at the roots of weeds in muddy ponds.

It need never be mistaken for a *Planorbis*, the general form of which it resembles, as it is furnished with an operculum. Its mouth, too, is perfectly circular, which is the case with no species of our English *Planorbis*.

Order *II.*—Pulmonobranchiata.

Family.—Limnæidæ.

Genus I.—Planórbis.

1. P. lineátus (*streaked*).

Quoit-shaped, compressed, more convex above than below, thin, semi-transparent, brownish or reddish horn-colour, carinated; *whorls* four; *spire* sunk; *umbilicus* narrow, but deep. Its special characteristic is the nautilus-like septa which are visible from the outside of the shell at intervals across the whorls.

This interesting species is local; it is found in sluggish streams and ponds in several parts of England, and also in Ireland.

2. P. nítidus (*shining*).

Quoit-shaped, much depressed, thin and glossy, yellowish or reddish horn-colour, sharply keeled; *whorls* four to five; *spire* sunk, but not so much as that of the last species; *umbilicus* small and shallow.

P. nitidus may be distinguished from *P. lineatus* in having no septa, being more depressed, the spire not so deeply sunk, in being more sharply keeled, and in having a shallower *umbilicus*.

It is a common species, inhabiting ponds and ditches on weeds in most parts of the country. It is frequently found covered with dirt, which is often extremely hard to remove.

3. P. NAUTÍLEUS (*like a nautilus*).

Quoit-shaped, having the upper side flat and the under side convex; dirty white or brown; striated in the line of growth by ridges of epidermis; *whorls* three.

There is no mistaking this elegant little shell, which is found on weeds and on the underside of decaying leaves in ponds and ditches throughout the country.

The variety, which is commonly found with the type, is an exceedingly beautiful object under a lens.

Var. *cristata.* Ridges of epidermis exaggerated to points.

4. P. ÁLBUS (*white*).

Flattish above, with *spire* depressed; dull white, very finely striated in the line of growth, but more distinctly marked with raised striæ spirally; microscopically hispid; *whorls* five.

This little shell often belies its name, being frequently found black with dirt. Under a powerful lens its delicate striations are visible, as well as rows of minute hairs running in a spiral direction.

It is a common species, found as far north as Aberdeenshire.

Var. *Draparnaldi.* Striæ in line of growth stronger.

5. P. GLÁBER (*smooth*).

Convex above, with a depression in the centre; concave underneath, rather thin, glossy, horn-colour, finely striated transversely, and still more faintly spirally; *whorls* five; *suture* very deep; *mouth* nearly circular; *umbilicus* large.

This is a very local shell, being only found in a few places in England, among which are Northumberland, Durham, Somerset, Norwich, and Birmingham. It resembles *P. albus* in shape, but is smooth and glossy. It is more liable, perhaps, to be taken for the young of *P. complanatus,* as the young of that species is only very faintly keeled.

6. P. SPIRÓRBIS (*circular coil*).

Very flat, rather solid, glossy, brown horn-colour; *whorls* five to six; faintly carinated on the lower margin.

This is a common shell throughout the British Isles. It is abundant on pond vegetation, and is often found eroded by infusoria, or with a black coating of mud, which is very hard to remove.

Var. *ecarinata.* Smaller, grey, no keel, only four whorls.

7. P. VÓRTEX (*whirlpool*).

Very flat, glossy, yellowish horn-colour; *whorls* six to eight; sharply keeled.

Found in similar habitats to the last species, but is less common. I have noticed that it is much more often found with a perfect epidermis than *P. spirorbis*, even in the same pond as that species.

It may be distinguished from *P. spirorbis* by its prominent keel; it is also flatter and thinner.

Var. *compressa.* Thinner and much flatter, and more sharply keeled.

8. P. CARINÁTUS (*keeled*).

Compressed, *spire* sunk, only slightly convex beneath, slightly glossy, semi-transparent, pale brown; *whorls* five to six; *suture* deep; *mouth* angulated above and below; sharply keeled in the *centre* of the outer margin.

Found in the south and east of England, in Lancashire, south of Scotland, in Wales, and Ireland; but it is local, and not plentiful.

Var. *disciformis.* Flatter and thinner.

9. P. COMPLANÁTUS (*flattened*).

Concave above, flat beneath, nearly opaque, strongly keeled below; *whorls* five to six; *mouth* rhomboidal.

This species is often so like the last that a very

natural doubt exists as to whether there are two distinct species. The typical difference is in the position of their keels, that of *P. carinatus* being in the centre of the outer margin, while that of *P. complanatus* is much lower—in fact, touching a flat surface when placed upon it. The typical *P. complanatus* is thicker and larger than *P. carinatus*, but it is difficult to decide upon half the individuals met with. The well-marked *complanatus*, however, is more often met with than the well-marked *carinatus*, and the former species is therefore allowed to be the commoner.

Var. I. *rhombea*. Smaller, more solid, more convex above and concave below. Only faintly carinated.

Var. II. *albida*. Whitish.

10. P. CÓRNEUS (*horn-coloured*).

Very tumid, *spire* sunk, opaque, dark horn-colour, lighter below; *mouth* nearly circular; *whorls* five to six.

This species is unmistakable, both from its size and shape. It is far the largest of our English *Planorbis*. When young it is covered with spiral rows of small hairs. It is very abundant in many districts in canals, streams, and ponds, but it is considered local. I have some specimens with four marks showing stages of growth, which, if they are formed only once a season, imply that these individuals were in their fifth year.

Var. *albina.* White, not uncommon.

11. P. CONTÓRTUS (*twisted up*).

Flat above, very convex below, dull brown; *whorls* eight, compressed; *suture* deep; *mouth* crescent-shaped; *umbilicus* deep.

This is a compact-looking little shell, and when once seen is never confounded with any other. It is fairly common on weeds in ponds and ditches.

Var. *albida.* Whitish.

10. P. DILATÁTUS (*expanded*).

Flattish above, extremely convex beneath; *whorls* two to two and a half, dull; nearly opaque; *mouth* exceedingly expanded, and very large; *suture* distinct; *umbilicus* narrow, but deep.

This little shell was imported from America in cotton bales. It was first noticed by Mr. T. Rogers, of Manchester, in 1869. The only two places where it has yet been observed to have made itself at home are the Bolton Canal and the Reddish Canal (between Stockport and Manchester), where there are cotton mills.

Genus II.—PHÝSA.

1. P. HYPNÓRUM (*frequenting the Hypnum, a moss*).

Sinistral, spindle-shaped, thin, highly glossy, semi-transparent, dark reddish horn-colour; *whorls* six to seven.

This pretty little shell is found in ponds and ditches widely distributed, but it is said to be local. It forms an attractive object in an aquarium as it floats on the top of the water with its dark blue foot uppermost.

2. P. FONTINÁLIS (*frequenting springs*).

Sinistral, oval, very thin, glossy, semi-transparent, greenish horn-colour; *whorls* four to five, tumid; *spire* very short, but pointed obtusely.

This shell, which is very common in running water and ponds on weed, is often disappointing when the animal is removed. When wet with the animal inside the shell appears iridescent, but when clean and dry is often dull.

This species is perhaps more interesting as an inmate of an aquarium than the last, as it is unusually active for a mollusk, and amusingly quarrelsome. When two individuals meet, they engage in (seemingly) angry combat, jerking themselves and their antagonist to and fro like wrestlers! They spin colourless threads of mucus, along which they crawl and practise gymnastics.

Var. I. *inflata.* Much larger.

Var. II. *curta.* Spire very short.

Var. III. *oblonga.* Spire produced.

Var. IV. *albina.* Milk white.

Genus III.—LIMNÆA.

1. L. GLUTINÓSA (*glutinous*).

Oval, globular, very thin and glossy, transparent, pale amber or greyish horn-colour, here and there striated in the line of growth, and faintly, but closely, striated spirally; *whorls* three to four, tumid; *spire* slightly produced; *suture* deepish; *mouth* nearly oval, contracted above.

This is a very local species, though abundant where it occurs. It is found in the home and eastern counties; also in Lake Windermere, Henley-on-Thames, Lake Bala, &c.

Var. *mucronata.* Spire produced.

2. L. INVOLÚTA ([*spire*] *enveloped* [*by whorls*]).

Oval, very thin, glossy, transparent, pale amber; *whorls* three to four, convex; *spire* sunk; *apex* distinct; *mouth* pear-shaped.

This is a beautiful delicately-formed shell, and it is to be regretted that it is only found in a small lake on the Cromlaun Mountain at Killarney. It is probably a variety of *L. peregra.*

3. L. PÉREGRA (*wandering*).

Obliquely ovate, thin, moderately glossy, semi-transparent, yellowish horn-colour; *whorls* five; *spire* raised; *apex* sharp; *mouth* large, pear-shaped.

This is the commonest and most variable of our freshwater shells, and is distributed over the whole of the British Isles.

Like other water-snails, *L. peregra* is amphibious, and may be found crawling on mud and stones uncovered by the water.

It is always worth the collector's while to gather specimens from any and every place he visits, and carefully label them in separate trays in his collection. Not till this is done will he have any idea of the variability of this species.

Var. I. *Burnetti.* Globular and solid, dull yellowish brown, strongly striated, last whorl nearly covering the rest, spire very short and blunt.

Var. II. *lacustris.* Somewhat resembling Var. I., but smaller and more glossy, with strong transverse grooves.

Var. III. *lutea.* Very solid, spire very short.

Var. IV. *ovata.* Ampullaceous, thin, whorls very convex, spire short.

Var. V. *acuminata.* Like the last, with more produced spire.

Var. VI. *intermedia.* Spire produced, mouth expanded.

Var. VII. *oblonga*. Oblong, compressed in front.

Var. VIII. *labiosa*. Smaller, outer lip expanded and reflected.

Var. IX. *picta*. Banded with white.

Var. X. *maritima*. Dwarfed, solid, spire produced, suture deep.

Var. XI. *succineæformis*. Shaped like a *Succinea*, very thin, whorls four.

Var. XII. *decollata*. Spire truncated.

Var. XIII. *sinistrorsa*. Sinistral, solid, spiral ridges distinct.

Var. XIV. *scalariformis*. Oblong, whorls somewhat disjoined.

4. L. AURICULÁRIA (*ear-shaped*).

Globosely ovate, thicker and paler than *L. peregra*, fairly glossy, semi-transparent; *whorls* four to five, very tumid; *spire* very short; *apex* sharp; *mouth* very large, with outer lip widely reflected.

This interesting species, named after its resemblance to the human ear, is found in rivers, lakes, ponds, and canals in many parts of the country. It is common and fine in the Thames, in canals in Lancashire, and other places. A soft tooth-brush may be used with great advantage to remove the dirt and confervoid which become very firmly attached to the shell.

Var. I. *acuta.* Smaller, more oblong, mouth narrower.

Var. II. *albida.* Smaller, thinner, and white.

5. L. STAGNÁLIS (*inhabiting swamps*).

Elongated, rather thick, not very glossy, pale whitish horn-colour, striated in line of growth ; *whorls* six to eight, body whorl forming most of the shell; *spire* tapering ; *mouth* large.

This is the largest member of the genus. It is generally distributed throughout the British Isles, inhabiting slow streams, canals, and standing water. This species also is often much improved by a careful brushing.

Var. I. *fragilis.* Smaller, more slender.

Var. II. *albida.* Shaped like Var. I., but white in colour.

Var. III. *labiata.* Dwarfed, solid, outer lip reflected and thickened.

6. L. PALÚSTRIS (*inhabiting marshes*).

Tapering, rather solid, dull brown ; *whorls* six to seven, tumid ; *mouth* pear-shaped.

This shell is found in ponds and slow running water throughout the British Isles. It is exceedingly variable both in size and form. I have some of the Var. *corvus*, which I took in Essex, one and a half inches in length, of a dark purple colour—truly magnificent shells.

Var. I. *corvus.* Larger, more tumid, dark purple.

Var. II. *elongata.* Spire more produced.

Var. III. *tincta.* Shorter and broader, mouth purplish.

Var. IV. *conica.* Conical, whitish, suture deep, having an umbilical cleft.

Var. V. *roseo-labiata.* With rose-coloured rib.

Var. VI. *decollata.* Spire truncate.

Var. VII. *albida.* White.

7. L. TRUNCÁTULA (*somewhat truncated*).

Somewhat resembling *L. palustris* in shape, but turreted in appearance with deeper *suture;* rather glossy; *mouth* ovate; *umbilical* cleft distinct.

This species is very liable to be confounded with some variety of the last, which in general form it somewhat resembles. It is, however, much smaller and more elegant in shape, with a much deeper suture, which gives it its turreted appearance, and more tumid whorls.

It is common both in still and running water, especially in shallow pools, which are often dried up in summer. It often climbs damp walls to a height of two or three feet. It varies considerably.

Var. I. *major.* Larger, more tumid.

Var. II. *elegans.* Larger, solid, slender.

Var. III. *minor.* Smaller, thin, dark horn-colour.

Var. IV. *albida.* Smaller, milk white.

Var. V. *scalariformis.* Smaller, whorls somewhat disjoined.

Var. VI. *microstoma.* Small, whorls swollen, mouth contracted.

8. L. GLÁBRA (*smooth*).

Cylindrical, tapering gradually to a blunt point, thin, glossy, greyish horn-colour, striated closely and regularly in line of growth; *whorls* seven to eight, rounded; *mouth* small for the size of the shell.

This is a local species inhabiting ponds and ditches in Yorkshire, Lancashire, Cheshire, and other parts of the country. I have often found them crawling up the stalks of reeds several inches out of the water. Some that I kept in an aquarium crawled up the glass side and tumbled out, thus falling victims to their adventurous disposition.

Var. *elongata.* Spire more produced.

Genus IV.—Áncylus.

1. A. FLUVIÁTILIS (*inhabiting rivers*).

Much the shape of the common marine limpet; *apex* recurved, giving the name *Ancylus* (little hook) to the genus.

Found very commonly adhering to stones or wooden piles in streams, but not in still water. It is common throughout Great Britain and Ireland, but is not found further north than Aberdeenshire.

Var. I. *capuloides*. Larger and higher.

Var. II. *gibbosa*. Smaller, with the hook overhanging the posterior margin.

Var. III. *albida*. White, more finely striated.

2. A. LACÚSTRIS (*inhabiting ponds*).

Compressed, oblong, thin; *apex* acute, interior often iridescent.

Found in sluggish streams, canals, ponds, &c., on the stalks of many aquatic plants, especially the *Stratiotes* and rushes. It is more local than the last species, but is abundant where it occurs. Its compressed, oblong shape marks it off distinctly from *A. fluviatilis*.

Var. I. *compressa*. Broader and flatter.

Var. II. *albida*. Milk white.

PART II.

TERRESTRIAL.

Order.—PULMONOBRANCHIATA (*continued*).

Family I.—LIMACIDÆ (*Slugs*).

The bodies of the *Limacidæ* are straight, not coiled; the upper part of the back is covered with a fleshy *shield* or *mantle*, under which the shell, if any, is located. This shell is rudimentary, and is usually an oblong plate, more or less membranaceous. They have four *tentacles* similar to those of snails, the longer or upper pair terminating in bulbs in which the eyes are situated. The whole lower surface of the body forms the *foot*. They have a respiratory orifice situated on the right side of the body by the edge of the mantle.

Genus I.—ARÍON.

All the *Arions* have a *slime gland* at the extremity of the tail.

1. A. ÁTER (*black*).

Body rounded in front, tapering behind; colour exceedingly variable, black, chocolate, reddish, yellowish, greenish, and whitish; coarsely tuberculated; *mantle* finely shagreened; *shell* usually absent, though occasionally represented by a few granules.

The great variations in the colour of this common slug do not appear to have much, if anything, to do with its habitat. The common type, coal black, from which it derives its specific name, is found in the same localities as beautiful cream-coloured specimens with dark tentacles and a yellow fringe. In Berkshire this variety is common, as also chocolate-coloured specimens. In the Isle of Man I once found a piebald specimen. In Ireland I have met with individuals of a bright yellow, and some of a gorgeous pink tint.

A few granules, only found in some specimens, represent the aborted shell. In chalky districts, however, some have a thick lump of nacreous matter under the mantle.

2. A. FLÁVUS (*yellow*).

This species is not so large as the last, averaging three inches in length, but the tubercles of the body are larger in proportion. It is of a yellow colour and has a wide range. There is no shell.

3. A. HORTÉNSIS (*inhabiting gardens*).

This slug is much smaller than either of the two last
species, measuring one and a half inches in length.   It
is usually slate grey, striped with black longitudinally.
It is "sluggish" to the last degree, being generally
found in a state of repose under stones.   The shell
consists of a few granules.

*Genus II.—*GEOMÁLCUS.

G. MACULÓSUS (*spotted*).

This is a very beautiful animal, only found in the
south-west of Ireland.   It is described as resembling
an *Arion* in form, two inches in length, glossy black
or brown with yellow spots, head grey, foot brownish
yellow with transverse stripes, shell oval and solid
with concentric lines of growth.

Mr. Rogers, of Manchester, informs me that he once
kept one of these slugs alive for more than two years,
in both of which it produced fertile eggs ; the young
ones, however, died when they had attained about half
their full size.

Genus III.—LÍMAX.

None of the members of this genus are furnished with a *slime gland.*

The members of this genus differ from the *Arions* in having their respiratory orifice nearer the posterior margin of the mantle.

1. L. GAGÁTES (*jet*).

Almost as varied in colour as *A. ater;* its characteristic distinction is in the mantle being bilobed when the animal is extended. Length two and a half inches. It is found in many parts of the country among vegetable matter like other slugs, but it is local.

2. L. MARGINÁTUS (*margined*).

This mollusk is characterized by its strong carination from the mantle to the end of the tail. It is of various colours—yellowish, rufous brown, or, more usually, pale slate grey; two and a half inches in length. The shell is remarkable for its thickness. I have taken some in the Isle of Man nearly cubical in form. It is a common species on stone walls in the country, and is very active after rain.

3. L. FLÁVUS (*yellow*).

This is a large species, attaining a length of four inches; yellowish, speckled with black and white so as to form a network. It is covered with coarse oval tubercles. The head and tentacles are bluish; the slime yellow. The shell is characteristic, having the nucleus slightly projecting over one end. It is found in cellars and damp places throughout the country.

4. L. AGRÉSTIS (*inhabiting fields*).

This is the well-known pest of London gardens, where it is usually creamy white. Its common aspect is white, mottled with dark grey. It has a thin oval shell, with a membranaceous margin. Length one and a half inches.

5. L. LÉVIS (*smooth*).

This is an active little creature, less than an inch in length, slender, very glossy, dark brown, mantle swelling behind and raised into a hump by the very solid shell. The shell is unguiform, very convex above and flat beneath, solid, nucleus terminal, margin sharp, and not membranaceous. It is found in the same situations as the rest of its tribe in many parts of the country.

6. L. TENÉLLUS (*tender*).

About three-quarters of an inch long, slender, shell oval, moderately thick.

This slug has been found in Shetland and North-umberland.

7. L. ÁRBORUM (*inhabiting trees*).

L. arborum varies very much in colour, and is some-times very beautifully marked. Length three inches. Its distinctive feature is the blunt point of the mantle behind. The shell is nearly oval, very thin and glossy. The nucleus is nearly terminal, the margin broad, and membranaceous.

It derives its name from its habit of climbing trees, a habit which, in common with members of the animal kingdom of a much higher development, it particularly affects in its youth. It is often found disporting itself by a thread of mucus which is attached to a branch.

Though found in most parts of Great Britain and Ireland, it is not abundant everywhere.

8. L. MÁXIMUS (*largest*).

This is a very handsome animal, mottled all over

with variable colouring. Adult specimens range from four to six inches in length. The shell is oblong, slightly convex above and concave below, marked with lines of growth; margin membranaceous.

These slugs congregate in holes in walls and in cellars. They come out at night to feed. The species is generally distributed.

Family II.—TESTACELLIDÆ.

Genus.—TESTACÉLLA.

1. T. HALIOTÍDEA (*resembling the Venus' ear shell, Haliotis*).

This slug is peculiar in having its shell nearly at the end of its tail, where its apology for a mantle is situated, almost covered by the shell.

It is carnivorous, eating earthworms, and possibly its own species. It is found at Norwich, Plymouth, Clifton, near London, and in the south of Ireland.

Var. *scutulum.* Shell narrower, more pointed.

2. T. MAÚGEI.

This species is found in nursery gardens at Bristol, whither it was doubtless brought with earth and foreign plants. The shell is larger and more cylindrical than that of *T. haliotidea.*

Family III.—Helicidæ.

Genus I.—Succínea.

1. S. PÚTRIS (*stinking*).

Shell pear-shaped, thin, nearly transparent, glossy, amber colour; *whorls* three to four; *spire* short; *mouth* long and open.

This species is found in moist places by streams and ponds crawling on plants. It particularly affects flags and withies. The early autumn is the time when it is most abundant. Though amphibious it is seldom found in the water. Sometimes colourless specimens are met with.

Var. I. *subglobosa.* Shorter, broader, and more solid.

Var. II. *vitrea.* Nearly membranaceous.

Var. III. *solidula.* Thicker, reddish yellow.

2. S. ÉLEGANS (*elegant*).

More slender than *S. putris*, rather thicker, mouth narrower.

This species appears to merge into the last, and until

E

lately they were regarded as doubtfully specifically distinct. Recent research, however, has shown a difference in the lingual dentition of the two species.

The main external difference between them is the more slender shape of the *elegans;* besides which, the suture close to the mouth is much deeper.

Both are found together in the same localities, and are equally common.

Var. I. *minor.* Smaller, thinner, reddish brown.

Var. II. *ochracea.* Smaller, thicker, larger spire, and smaller mouth.

3. S. OBLÓNGA (*oblong*).

Shaped somewhat like *L. truncatula*, rather more solid than the two last species; brownish yellow horn-colour; *whorls* three to four, convex; *suture* deep.

This is a rare species, found near the sea-coast at Swansea, North Devon, Glasgow, and Cork.

It is very distinct from the other two species, being very much smaller, with a longer spire in proportion to its size, and a very deep suture.

Genus II.—VÍTRINA.

V. PELLÚCIDA (*transparent*).

Spheroidal, very thin, transparent, green; *spire* scarcely raised; *mouth* large.

The *Vitrinas* are a link between the slugs and snails proper. They have the same tooth formation and mantle as the slugs, while their shells resemble those of the *Helices*.

Our pretty little English species is found in almost all parts of the British Isles. This shell, like a bubble of clear greenish-tinted glass, is a truly beautiful object, especially when set out on white cardboard or with a background of cotton wool. Two varieties are found, but the shell is very uniform. I have, however, a specimen taken in the north of Ireland with two milk-white bands running in a spiral direction from apex to mouth.

Under a lens, or even with the naked eye, the operation of breathing can be well observed. In autumn, winter, and early spring, especially after rain, our little friend is abroad on dead leaves and among moss.

Var. I. *depressiuscula*. Oval and flatter on both sides.

Var. II. *Dilwynii*. Globular, spire produced.

Genus III.—ZONÍTES.

The identification of the different species in this genus gives more trouble to the novice—to whom they all look pretty much alike—than all the rest of British land and freshwater shells put together, with the exception of the minute *Vertigos*. When the adults of two species differ considerably in size, as *radiatulus* and *nitidus*, and the markings and sculpture are much alike, it is a good plan to *count the whorls* of a small specimen. A full-sized *radiatulus* has four and a half, and *nitidus* five. If, therefore, the shell under consideration is the size of *radiatulus*, and has three whorls, the observer will know that another whorl and a half would make the shell too large for a *radiatulus*, and he may conclude it to be a young *nitidus*.

The *animals* should be carefully compared. Of the ten species, *fulvus, crystallinus, purus, radiatulus,* and *excavatus* are unmistakable when their striking differences have been once pointed out.

1. Z. CELLÁRIUS (*inhabiting cellars*).

Compressed, glossy, semi-transparent, pale horn-colour above, opaque white below; *spire* nearly flat; *mouth* deeply semi-lunar; *whorls* five to six; *umbilicus* broad and deep; *diameter* half an inch.

This is the largest of the English *Zonites.* It is common everywhere in damp cellars, under stones and logs.

Var. I. *complanata.* Rather smaller, spire very flat.

Var. II. *albida.* White or colourless.

Var. III. *compacta.* Spire more raised, whorls more convex.

2. Z. ALLIÁRIUS (*smelling of garlic*).

More convex above, though less so below, than *Z. cellarius; whorls* five, the last not so large in proportion as that of the last species; *spire* slightly produced; *mouth* narrow; *umbilicus* rather wider in proportion than that of the last species; *diameter* a quarter of an inch.

Z. alliarius is rather more local than *Z. cellarius,* though equally distributed. It has a powerful odour of garlic, but this is not a certain mark of identity, as other species have the same peculiarity. In the north of Ireland, near Coleraine, the variety is very abundant under stones, but I have rarely come across the typical form in that district.

The umbilicus is wider in proportion to the size of the shell than that of the last species, and the spire is more produced.

Var. *viridula.* Greenish white.

3. Z. GLÁBER (*smooth*).

Convex above and below, very glossy, semi-transparent, pale horn-colour, only slightly marked with white below; *spire* raised, and produced to a point; outline of *mouth* forming about three-quarters of an almost symmetrical circle; *whorls* five to six; *umbilicus* deep, but not disclosing so much of the last whorl as that of *Z. cellarius; diameter* slightly more than three-eighths of an inch.

This shell, which is distributed widely throughout England, and has also been found in Wales and the south of Scotland, is local.

Its discovery in this country is due to Mr. Thomas Rogers, of Manchester, who first noticed it in 1870.

It is likely to be mistaken for *Z. cellarius,* from which it differs in the following respects :—

(1.) It measures less in diameter.

(2.) It measures more in height in proportion to its diameter.

(3.) The spire is raised by a slight inclination to a point.

(4.) The outline of the mouth is nearly circular instead of oval.

(5.) The umbilicus does not disclose the penultimate whorl, as is the case with *Z. cellarius.*

(6.) When the two shells are viewed directly towards the mouth, the interior of the mouth of *Z. cellarius* is seen to be more interrupted by the penultimate whorl.

(7.) It has less white below, and is altogether slightly more smooth and transparent.

In shape it more nearly resembles *Z. alliarius*, but its size is sufficient to prevent its being mistaken for that species.

Mr. Rogers tells me that when the animal is within the shell it may easily be recognized by the abrupt termination of the colour of the mantle, which is distinctly visible through the transparent shell.

4. Z. NITÍDULUS (*slightly glossy*).

Compressed, thin, only slightly- glossy, whitish underneath; *whorls* five, convex; *spire* slightly raised; *suture* rather deep; *mouth* round, except where interrupted by the last whorl; *umbilicus* very broad and deep; *diameter* between a quarter and three-eighths of an inch.

This shell is widely distributed, and, like the rest of the genus, frequents moist places under stones and in woods.

It may be distinguished from *Z. cellarius* by its smaller size and its much larger umbilicus. It has, moreover, one whorl less, and is not nearly so glossy.

Var. I. *nitens*. Smaller and lighter, last whorl very large.

Var. II. *Helmii.* Like Var. I., but pearl white.

5. Z. PÚRUS (*clear*).

Compressed, extremely thin, dull, semi-transparent, light brown horn-colour; very faintly striated in both directions, and having consequently a reticulated appearance under a microscope; *whorls* four, the last very large ; *mouth* almost circular; *umbilicus* rather narrow, but deep; *diameter* nearly one-seventh of an inch.

Found in many parts of Great Britain and Ireland, but not everywhere. It is a very fragile shell, and the animal is difficult to extract entire.

The variety is an exquisitely delicate object.

The animal is cream-coloured.

Var. *margaritacea.* Pearl white and nearly transparent.

6. Z. RADIÁTULUS (*slightly rayed*).

In size and shape resembling the last species, but highly glossy; darker in colour, and exquisitely marked by regular curved striæ in the line of growth—more strongly above than below; *whorls* four and a half; *diameter* nearly one-seventh of an inch.

Dr. Jeffreys says that this species frequents moister places than *Z. purus.* It is not abundant though widely distributed in Great Britain and Ireland. Under a lens it is a beautiful object, and when once seen need never be mistaken for the last species, but care should be taken not to confound it with the young of *Z. nitidus.*

Var. *viridescenti-alba.* Greenish white.

7. Z. NÍTIDUS (*glossy*).

Rather solid, glossy, and semi-transparent, chocolate brown, transversely striated, faintly granulated under a microscope; *whorls* five; *spire* raised, apex blunt; *suture* deep; *mouth* round; *umbilicus* narrow, but deep; *diameter* a quarter of an inch.

It is found at the roots of grass in moist places, not very abundantly. Its most distinctive characteristic is the absence of white on the under side.

Var. *albida.* White.

8. Z. EXCAVÁTUS (*hollowed out*).

Not very glossy, semi-transparent, light brown, coarsely striated; *whorls* five and a half, nearly cylindrical; *spire* rather prominent; *suture* very deep; *mouth* nearly circular; *umbilicus* very broad and deep; *diameter* nearly a quarter of an inch.

This is a well-marked species. Its broad and deep umbilicus, its convex spire, cylindrical whorls, and coarse appearance combine to render it easy of identification. The young shells are slightly carinated.

It is found in various parts of England, Scotland, Ireland, and Wales, but it is a local shell. Not far

from Manchester, on the borders of Cheshire, both it
and its variety are not uncommon under stones, and
along canal banks among the roots of grass in the same
neighbourhood.

Var. *vitrina.* Transparent, greenish white.

9. Z. CRYSTÁLLINUS (*crystalline*).

Depressed above, more convex below, thin transparent
pearly white; *whorls* five, compressed; *mouth* semi-lunar;
umbilicus exceedingly small; *diameter* one-eighth of an inch.

The iridescent appearance and symmetrical form
of this little shell cause us to regret that it is one of
the smaller instead of the larger members of the genus.

It is easily identified from its colour and the compressed appearance of the whorls, but especially by its
umbilicus, which is so small that a pin of moderate
size will hardly enter.

It is common everywhere in moist sheltered situations.

Var. *complanata.* Nearly flat on both sides, the
last whorl larger.

10. Z. FÚLVUS (*tawny*).

Pyramidal, thin, glossy, of a clear semi-transparent
horn-colour; *whorls* five and a half, cylindrical; *mouth*
narrow; *umbilicus* merely a slight indent, only found on
adult specimens; *diameter* one-tenth of an inch.

This pretty little shell is a link between the *Zonites* and *Helices*. Its quaint pyramidal shape prevents a possibility of mistaking its identity.

Though generally distributed, it is by no means common everywhere. It should be sought under rotten branches and logs and among moss in woods.

Var. *Mortoni*. Paler, spire more depressed, more sharply carinated.

Genus IV.—HÉLIX.

1. H. LAMELLÁTA (*having plates*).

Globosely conic, semi-transparent horn-colour, epidermis raised into plaits in the line of growth; *whorls* six; *spire* blunt; *umbilicus* distinct.

This little shell, which inhabits the north of Great Britain and most parts of Ireland, as well as Anglesea, frequents dead leaves, especially those of the holly.

It somewhat resembles an immature *Pupa* in shape, but a second glance will show its distinct umbilicus and its plaits of raised epidermis, which form regular ridges in the line of growth.

2. H. ACULEÁTA (*spiny*).

Pyramidal, turreted in appearance owing to its rapidly increasing *whorls*, of which there are four or four and a

half. As in the last species the epidermis is raised into
ridges in the line of growth, but in this shell the ridges are
exaggerated into points.

This is a beautiful object under a lens. It is pretty
generally distributed, and is found among moss and
dead leaves in moist shady situations. It may often
be passed over as a small lump of dirt owing to its
colour and shape.

Var. *albida.* Colourless.

3. H. POMÁTIA (*operculate*).

Globose, solid, cream-colour banded with red, coarsely
striated in the line of growth; *umbilicus* narrow.

This fine shell is the largest of our snails. Though
common on the Continent, it is only sparingly met
with in a few localities in the south of England. It
derives its name from a Greek word (poma) signify-
ing an operculum, on account of the thick epiphragm
which it makes on retiring into winter quarters. This
epiphragm is formed by a secretion of the animal, and
hardens on exposure to the air like plaster-of-Paris.
When this thick plate closes the entrance, the animal
retires deeper into the shell, and fortifies itself still
further against the cold by a succession of thin films.

The notion that this shell was introduced by the

Romans as an article of food is now discarded. On the Continent it still continues to be considered a delicacy.

Var. *albida.* White.

4. H. ASPÉRSA (*besprinkled*).

Globose, solid, opaque, yellowish with dark brown bands, which are interrupted at intervals by lighter markings; *whorls* five and a half; *umbilicus* none.

Everyone knows the "common garden snail." Those, however, who only regard it as an object of immediate extermination have no idea of its beauties when met with under happier auspices.

The proprietor of a London garden, for instance, strolling out after a shower of rain, when he fixes his stern gaze upon our friends regaling themselves upon his pet sunflowers, generally collects them in an empty flower-pot, and then heaves them stealthily into his neighbour's garden, whence, perhaps the next day, they return in a similar manner. This person, I say, is not in the frame of mind to appreciate their beauty; and, indeed, near towns they are *not* beautiful, being mostly without any epidermis and covered with dirt.

The finest and most beautifully marked specimens that I have seen were taken at the Isle of Man. In various parts of England *H. aspersa* still forms an article of food, and in Venice and other parts of the Continent I have seen baskets-full collected for this purpose. The taste is insipid, and the animals are apt to be tough if not well boiled. As nourishing food, however, they rank with calf's-foot jelly, oysters, &c.

A sinistral monstrosity occasionally occurs which is said to be worth a guinea.

Var. I. *albo-fasciata*. Reddish brown, with a single white band.

Var. II. *exalbida.* Yellowish or whitish.

Var. III. *conoidea*. Small, thin, conical (sand-hills on sea coast).

Var. IV. *tenuis*. Dwarfed, thin (Channel Isles).

5. H. NEMORÁLIS (*inhabiting groves*).

Globular, solid, very brilliantly and variously coloured; *whorls* five and a half; *umbilicus* distinct in the young, but covered in the adult; inside lip reddish brown.

This is the common and pretty snail which spots the country hedges after a shower of rain. It is also found in great abundance on sand-hills by the sea coast.

There are said to be nearly two hundred colour varieties of the common type, and nearly fifty of the Var. *hortensis!* The collector will do well to make a series of varieties according to his fancy, keeping three or four of each sort, and then to fill up the gaps with specimens that are intermediate in appearance. In course of time he will come to the conclusion that no distinct line can be drawn between any of them—a startling conclusion to anyone acquainted with only one or 'two distinct types—say the dark chocolate and the pure yellow.

Thrushes devour them with great relish, breaking the shells on a flat stone. I have observed fresh shells, empty and broken, *apparently* gnawed by rabbits' teeth, at the entrance to their burrows in the sand-hills ; but whether "Brer Rabbit" is actually to blame for this I have been unable to discover.

In Alsace I have observed these shells high up in trees, a habit which I have not noticed in England.

There seems some doubt if the different varieties interbreed. The collector should always examine the mouths of all pairing couples he comes across, and make notes of his observations.

Var. I. *hortensis.* Smaller, more globular, white lipped.

Var. II. *hybrida.* Same size as Var. I., mouth pink.

Var. III. *major.* Larger, thinner, more depressed.

Var. IV. *minor.* Dwarfed.

6. H. ARBUSTÓRUM (*living in copses*).

Globose, solid, glossy, brown and yellow, forming a sort

of check pattern, surrounded with a single band of dark brown; lip white and thick; *umbilicus* indistinct.

This is a rather local species, inhabiting dry ditches. Where it is found it is abundant. In some parts of Derbyshire the white variety occurs in great profusion with the type.

Var. I. *flavescens.* Yellowish white, without any circular band.

Var. II. *major.* Larger, more depressed.

Var. III. *alpestris.* Smaller, spire raised.

Var. IV. *fusca.* Dark brown, with or without the band, very thin.

7. H. Cantiána (*Kentish*).

Compressed, thin, dull, creamy white with a slight tinge of pink; closely striated in the line of growth; *whorls* six to seven, rounded, not keeled; *spire* short, obtuse; *mouth* pink inside; *umbilicus* narrow, but deep.

This shell is found in Northumberland, Yorkshire, Wales, and some of the southern counties of England. Where it occurs it is common.

When young it is covered with hairs, which afterwards disappear.

8. H. Cartusiána (*named after a Carthusian monastery*).

Sub-conic, depressed, thick, not very glossy, yellowish

white tinted brown, often marked with a whitish band; *whorls* six to seven, faintly keeled; *mouth* furnished with a broad white internal rib; *lip* thin; *umbilicus* almost hidden by the reflection of the outer lip.

This species is only found on the downs of Kent and Sussex, near the coast, where it is abundant.

It is much smaller and smoother than *H. Cantiana*.

9. H. RUFÉSCENS (*reddish*).

Sub-conic, depressed, thick, semi-transparent, dull brown, more or less reddish with a whitish line round the body whorl, which is bluntly keeled; *spire* short, obtuse; *mouth* semi-lunar, with internal white rib; *umbilicus* narrow, but deep.

This is a very common shell in woods, hedgerows, and under stones. The young are sometimes hispid. The varieties are moderately common.

Var. I. *albida*. White or colourless.

Var. II. *minor*. Smaller, spire more produced.

10. H. CONCÍNNA (*neat*).

Sub-conic, rather solid and glossy, light horn-colour, sometimes streaked with reddish brown. A whitish band is often found on the last whorl, which is faintly keeled. *Whorls* six and a half; *spire* obtuse; *suture* deep; *mouth* semi-lunar, with white internal rib; *umbilicus* rather broad and deep. Hairs, which are easily rubbed off, are found scattered scantily over the epidermis.

F

Conchologists have been much puzzled whether this shell is merely a variety of *H. hispida* or a distinct species. The two shells are found in similar situations, and are equally distributed.

The following are the distinctions made by Dr. Jeffreys :—

H. CONCINNA.	H. HISPIDA.
Never globose.	Globose.
Moderately glossy.	Very slightly glossy.
Umbilicus wide.	Umbilicus narrow.
Hairs scattered and deciduous.	Hairs thicker and permanent.
Animal darker, foot grey.	Foot thick and yellowish.

Var. I. *albida.* White.

Var. II. *minor.* Smaller, white, spire produced.

11. H. HÍSPIDA (*hairy*).

Sub-conic, thin, only slightly glossy; colour markings the same as *H. concinna*, but darker; *whorls* six and a half; *spire* blunt; *suture* deep; *umbilicus* narrow, but deep. Covered with short white hairs arranged in spiral lines.

Found under stones, timber, and among moss and leaves ; common and universally distributed. It is a pretty object when in full " plumage."

Var. I. *subrufa.* Reddish brown, solid, with a strong labial rib.

Var. II. *albida.* Thin, white.

Var. III. *conica.* Smaller, spire raised.

Var. IV. *nana.* Much smaller, depressed, with a strong labial rib.

Var. V. *subglobosa.* More globular, thinner, light coloured, umbilicus small.

12. H. SERÍCEA (*silky*).

Sub-globular, thin, rather glossy, very pale grey, some-times tinged with yellow; thickly covered with white silky hairs; *whorls* six, tumid; *spire* raised, obtuse; *suture* rather deep; *mouth* semi-lunar, with white internal rib; *umbilicus* very small.

This species is local, but abundant where it occurs. It is found in Scotland, Yorkshire, Anglesea, Hants, &c.

Its chief distinction from *H. hispida* is its umbilicus, which is very small, while that of the last species is broad and deep; besides which, the form of *H. sericea* is more globular, the spire more raised, and the bristles much thicker. The usual colour of *H. hispida* is much darker, though the Var. *subglobosa* is much like *H. sericea* both in shape and colour.

Var. *cornea.* Horn-colour, very thin, labial rib strong.

13. H. RELEVÁTA (*discovered*).

Somewhat globular; *spire* depressed; pale olive green, covered scantily with short pale hairs; *whorls* four to four and a half; *suture* deep; *mouth* nearly circular; *umbilicus* small.

This is a very local species, only found in the Channel Islands and on the south-west coast of England.

14. H. FÚSCA (*dusky brown*).

Sub-conical, so thin as to be almost membranaceous, glossy, pale yellowish brown, marked in the line of growth by strong corrugations; faintly keeled; *whorls* five to five and a half; *umbilicus* very small.

The shell of this species is almost membranaceous, and has been known to have been pressed flat and blown up again without suffering any detriment.

It does not occur everywhere, but is found near Belfast, in Wales, and in several parts of England.

It frequents grassy banks, and is said to feed on nettles.

15. H. PISÁNA (*first noticed at Pisa*).

Sub-globular, solid, nearly opaque, yellowish white, marked with variegated bands; *whorls* five to five and a half; interior of mouth rose-colour; *umbilicus* very small.

This species is unmistakable from its markings and shape, though it somewhat resembles a large form of

H. virgata. It is found on sand-hills near the sea in South Wales, Cornwall, Dublin, and Jersey.

In the south of France and in Italy it is collected in great quantities for food, and may be seen exposed for sale in the markets. In Marseilles and Venice I have seen this species in countless thousands adhering to nettles and other plants, apparently enjoying the noon-day sun.

Var. *alba.* White or pale yellow.

16. H. VIRGÁTA (*striped*).

Conical, globose, solid, almost opaque, cream-coloured, very variously banded; *whorls* six, tumid; *spire* raised; *mouth* nearly circular, furnished with an internal lip of a pinkish colour; *umbilicus* narrow, but deep.

This very variable shell is found on downs and sand-hills in many parts of the British Isles—but not in Scotland—especially near the sea coast. It is exceedingly abundant in most places where it occurs, and as it varies greatly in size, colour, and markings, a good series ought to be secured.

The young are carinated.

Var. I. *subaperta.* Whitish, depressed, umbilicus very wide.

Var. II. *subglobosa.* Smaller, with a double band above the periphery; umbilicus wide.

Var. III. *submaritima.* Smaller, darker, spire more raised.

Var. IV. *carinata.* Yellowish white, depressed, keeled.

Var. V. *leucozona.* Darkly banded, with a clear white band round the centre of the shell.

17. H. CAPERÁTA (*wrinkled*).

Sub-conical, compressed, solid, opaque, dull, dirty white, streaked and banded by variable lines which are sometimes merely a row of spots; obtusely carinated; *whorls* six; *mouth* nearly round, furnished with a white internal rib; *umbilicus* open and shallow.   •

Found under stones and on the stalks of grass on sandy soils, especially near the sea coast, where it is as abundant as the last species, and frequently found with it.

It is a particularly hardy animal, and will often be found feeding in winter when other snails are torpid. It has as great a power of enduring heat as cold, and I have observed it not only in many parts of southern Europe, but even on the desert in the neighbourhood of Bagdad.

It is, perhaps, apt to be confounded with the young of *H. virgata,* which is also carinated; but a very little experience will enable the collector to determine its identity. It is more depressed and has a larger umbilicus than the adult of the last species, and is smaller than the usual type.

Var. I. *major.* Larger.

Var. II. *ornata.* Smaller, with a broad, dark band.

Var. III. *subscalaris.* Spire more raised, whorls more tumid.

Var. IV. *Gigaxii.* Smaller, spire more depressed.

18. H. ERICETÓRUM (*inhabiting heaths*).

Almost circular, very flat, thin, nearly opaque, glossy, cream-colour, with one broad dark band above and several narrow ones beneath; *spire* slightly raised; *whorls* six, cylindrical; *mouth* nearly circular, often of a reddish colour inside; *umbilicus* very large and open.

Found in considerable quantities on heaths, downs, and sand-hills—particularly by the sea. It appears to be as fond of thistles as the donkey, but without that animal's reasonable excuse!

Var. I. *alba.* Milk white.

Var. II. *minor.* Much smaller, sometimes white, sometimes banded.

Var. III. *instabilis.* Smaller, spire more raised, darker, streaked or spotted.

19. H. ROTUNDÁTA (*rounded*).

Circular, compressed, thin, and semi-transparent, pale

horn-colour, marked at regular intervals by rufous brown; ribbed in the line of growth; *whorls* six to seven, compressed above; outer margin strongly carinated; *umbilicus* large and shallow.

This is an extremely common shell, and is easily identified. It is found everywhere, under stones, logs, &c., and in moss. The white variety, which is a rare capture, is a beautiful object.

Var. I. *minor.* Smaller.

Var. II. *pyramidalis.* Spire much produced.

Var. III. *Turtoni.* Spire very depressed.

Var. IV. *alba.* Cream-coloured or pale green.

20. H. RUPÉSTRIS (*inhabiting rocks*).

Sub-conical, rather solid, semi-transparent, rather glossy, dark horn-colour; closely and regularly striated in the line of growth; obtusely carinated, especially when young; *spire* raised; *suture* very deep; *mouth* gibbous; *umbilicus* wide and deep.

This is a hardy little creature, and may be found on stone walls and limestone rocks at a considerable elevation. It is gregarious.

Var. *viridescenti-alba.* Greenish white.

21. H. PYGMÆA (*minute*).

Circular, depressed, thin, semi-transparent, glossy,

yellowish horn-colour; striated finely and regularly in the line of growth; *whorls* four, cylindrical; *spire* more or less raised; *umbilicus* large.

This is the smallest of the *Helices*. It frequents moist situations in woods and under hedges among dead leaves. A good plan that saves much time is to take a quantity of dead leaves home, and after drying them, to examine the siftings.

H. pygmæa is a very pleasing object under a powerful lens—the close-set, well-marked striæ appearing as a surprise to the observer.

22. H. PULCHÉLLA (*minutely beautiful*).

Nearly circular, rather solid, semi-transparent, white; *whorls* three and a half; *mouth* nearly circular, with rim thickened, giving it a trumpet-shaped appearance; *umbilicus* wide and shallow.

This beautiful little shell is found under stones, among moss, and on sand-hills in many parts of Great Britain and Ireland. Its peculiarly shaped mouth is a sufficiently distinctive mark of its identity.

The variety is not uncommon.

At Beaumaris, where it is abundant, I have noticed that the type is comparatively rare.

Var. *costata*. Marked in line of growth with membranaceous ribs.

23. H. LAPICÍDA (*stone-cutter*).

Circular, compressed, dark rufous brown, dull, semi-transparent; *whorls* five; *mouth* obliquely oval, surrounded by a strong white reflected rim; very strongly carinated; *umbilicus* large.

This strikingly formed shell is found in many parts of England, rarely in Scotland and Wales, and not at all in Ireland. It occurs chiefly in calcareous districts, but not exclusively; it is fairly common at Maidenhead (Berks), where the soil is gravelly, and at other places.

The name "stone-cutter" is due to its supposed habit of boring into rocks. This idea is, of course, erroneous, but it does ensconce itself in crevices of rocks, whence it emerges after rain.

Var. I. *albina.* White.

Var. II. *minor.* Smaller and darker.

24. H. OBVOLÚTA (*wrapped up*).

Circular, flat above, compressed below, rather solid, opaque, reddish brown, hispid; *whorls* six and a half, cylindrical; *mouth* triangular, surrounded by a strong pinkish white rim with three tooth-like protuberances; *umbilicus* large.

This peculiar shell is only found in the neighbourhood of Ditcham Wood, Hants, where it is not uncommon. It is common on the Continent, whence it is thought possible that it was introduced.

25. H. VILLÓSA (*covered with hairs*).

Both in shape and size much like *H. rufescens;* of a pale semi-transparent white; *spire* flat; no carination; covered with hairs.

This shell—four specimens of which were found at Cardiff in 1873—has been admitted into the British list.

How it found its way thither has not been satisfactorily explained, but the neighbourhood of a large seaport suggests its introduction among goods of some sort.

Genus V.—BÚLIMUS.

1. B. ACÚTUS (*pointed*).

Conical, slightly diaphanous, whitish with irregular transverse streaks of a darker shade; *whorls* eight to nine, convex; *umbilicus* slight.

This very abundant species is only found on the coast on grassy cliffs, downs, and sand-hills. The finest specimens that I have ever met with are those I have taken on the sand-hills at Port Rush (N. Ireland). The varieties are equally common with the type.

Var. I. *bizona.* Two dark bands on the body whorl.

Var. II. *inflata.* More tumid, shorter, streaked with brown, or marked with a single band.

2. B. MONTÁNUS (*inhabiting mountains*).

Globosely conic, slightly glossy, light brown; *whorls* seven and a half; *spire* tapering, but blunt; *lip* white, reflected; *umbilicus* narrow, but deep. Immature specimens are keeled.

This is a local shell, being confined to the southern and western counties of England. It has a habit of ascending the ash and the beech in the spring, presumably to feed and pair, descending in the autumn to hibernate.

3. B. OBSCÚRUS (*hidden*).

Of the same shape as the last species, but much smaller, rather shorter in proportion, and more glossy; transparent horn-colour; *whorls* six and a half.

The name *obscurus* was given to this shell in consequence of its habit of covering itself, by means of its slime or an exudation of the epidermis, with earth or any substance it comes in contact with,—thus rendering itself inconspicuous. In some districts, where the extraneous matter is not suitable for attachment (as in the calcareous districts of Derbyshire), the shell is found clean and smooth in the crevices of rocks. Though often met with in hedge-banks among damp moss and earth, it never occurs in any quantities.

Var. *alba.* White or colourless.

Genus VI.—Púpa.

1. P. secále (*a grain of rye*).

Globosely conical, solid, opaque, rufous brown; regularly striate in the line of growth; *whorls* eight to nine; *mouth* horseshoe-shaped, furnished with eight or nine white ridges having the appearance of teeth.

There is no mistaking this species. It is far the largest of our *Pupæ.* Though local, it is abundant where it occurs—on rocks, in woods, chiefly in calcareous districts in England and South Wales, but not in Ireland or Scotland.

Var. I. *alba.* White.

Var. II. *Boileausiana.* "It is distinguished from the type by its smaller size, the larger plication being always double, and by the presence of a prominent additional fold at the angle of the columella."

Var. III. *edentula.* Smaller and thinner, smooth and glossy, tooth-like processes absent.

2. P. RÍNGENS (*grinning*).

Ovate, rather solid, glossy, dark horn-colour; *whorls* six and a half; *mouth* horseshoe-shaped, furnished with seven or eight tooth-like processes which, as well as the reflected lip, are often of a reddish tinge; *umbilicus* small, but distinct.

The name *ringens*, which signifies " grinning like a dog," *i.e.* " showing its teeth," was of course suggested by the appearance of the mouth, the denticular processes of which are a safe mark of distinction from the two following species.

It lives among moss and dead leaves in moist situations in Scotland, the north of England, and Ireland generally.

Young specimens are furnished with internal septa, which are visible from the outside, and are carinated.

Var. *pallida*.   Whitish.

3. P. UMBILICÁTA (*having an umbilicus*).

Sub-cylindrical, rather solid, glossy, semi-transparent, pale horn-colour; *whorls* six to seven; *mouth* horseshoe-shaped, with a broad strong white lip; on the base of the penultimate whorl there is sometimes a denticle; *umbilicus* small.

This is a very abundant and common species throughout the British Isles on stone walls and among moss, dead leaves, &c. As Dr. Jeffreys remarks, "the spire varies greatly in length," some individuals not being half the length of the full size, which is between one-seventh and one-eighth of an inch.

The young are carinated and have a well-marked umbilicus.

Var. I. *edentula.* Denticle absent.

Var. II. *alba.* White.

4. P. MARGINÁTA (*margined*).

Cylindrical, rather solid, dull, brown horn-colour; *whorls* six to seven; striated finely in the line of growth; *mouth* nearly circular, sometimes furnished with a denticle similar to that of the last species. *Outside* there is a white rib, not quite at the margin, which is thin and unreflected.

This species is specially fond of sandy soils near the sea coast, but is frequently found inland. Generally distributed.

P. marginata and *P. umbilicata* differ in the following respects:—

P. UMBILICATA.	P. MARGINATA.
Smooth and glossy.	Dull, striated.
Ovate.	Cylindrical.
Mouth horseshoe-shaped, with a thick reflected lip.	Mouth smaller, nearly circular, lip not reflected, but thickened.

Var. I. *bigranata*. Smaller, thicker, with two denticles.

Var. II. *albina*. White.

Genus VII.—VERTÍGO.

The members of this genus have only two tentacles, while those of the preceding genus have four. In order to aid the collector to identify these minute and very puzzling little shells, I have tabulated some of the most important differences of those which are dextral and furnished with teeth. (See Appendix B.)

The ten species have been conveniently arranged as follows :—

A. *dextral, denticulated.* ANTIVERTIGO, LILLJE-BORGII, MOULINSIANA, PYGMÆA, ALPESTRIS, SUB-STRIATA.

B. *sinistral.* PUSILLA, ANGUSTIOR.

C. *dextral, without teeth.* EDENTULA, MINUTIS-SIMA.

A. DEXTRAL, *denticulate.*

1. V. ANTIVERTÍGO (*not reversed*).

Oval, thin, semi-transparent, glossy, rufous horn-colour; *whorls* four and a half, tumid; *spire* short, abrupt; *mouth* semi-oval, contracted in the middle of the outer edge;

teeth six to nine, of a reddish brown colour; *umbilicus* distinct.

Found in moist places under logs, stones, among moss, and on water plants, as well as in elevated situations, in many parts of the British Isles.

Young specimens have only two denticles, one on the pillar and one at the base of the penultimate whorl.

2. V. LILLJEBORGII[1] (*after the Swedish naturalist,*
Lilljeborg).

Barrel-shaped, very thin and glossy, light horn-colour; *whorls* four and a half, very tumid; *spire* short, blunt; *suture* very deep; *mouth* semi-oval, constricted at the outer edge; *teeth* four; *umbilicus* distinct.

Dr. Jeffreys found this species in the west of Ireland, and figured and described it in his work under the name of *V. Moulinsiana,* from which it is doubtfully distinct.

As to the differences of this from the last species and *V. pygmæa* I cannot do better than quote Dr. Jeffreys, who says: "This species differs from *V. antivertigo* in being larger, more ventricose, and of a much lighter colour, in the mouth and outer lip not being contracted, and especially in the number and position of the teeth,

[1] Identical with *V. Moulinsiana* of Dr. Jeffreys' "British Conchology."

which never exceed four, instead of from six to ten, as in that species. From *V. pygmæa* it may be distinguished by being twice the size and very much more ventricose, and also of a lighter colour. The difference is equally great between all three species. *V. Lilljeborgii*[1] resembles *V. antivertigo* in form and *V. pygmæa* in the number of teeth. It is among the largest of our native species of *Vertigo*."

Var. *bidentata*. "Labial or palatal teeth wanting."

3. V. MOULINSIANA (*after the French conchologist, Des Moulins*).

Dr. Jeffreys, writing in the "Annals and Magazine of Nat. Hist." for Nov., 1878, says: "The shell of *V. Moulinsiana* is rather more swollen than that of *V. Lilljeborgii*, and the labial rib is much stouter." *V. Moulinsiana* has often an extra denticle.

V. Moulinsiana, if really specifically distinct from *V. Lilljeborgii*, approaches it very nearly in form. My figure is an enlargement of a specimen taken at Hitchin, which a friend was kind enough to lend me.

This rare shell has been found in Hertfordshire and also in Hampshire.

---

[1] Written *V. Moulinsiana* in "British Conchology.

4. V. PYGMÆA (*minute*).

Oval, somewhat cylindrical, semi-transparent, glossy, horn-colour; *whorls* four and a half, convex; *spire* blunt; *mouth* semi-oval; *teeth* four to five (one on the body, one on the columella, two or three inside the outer lip); *umbilicus* small, but deep.

This minute species is found in most parts of Great Britain and Ireland at the roots of grass, and under stones and logs in dry as well as moist situations.

It differs from the last two species in being much smaller and more cylindrical in form. The outer lip, moreover, is not constricted or angulated.

Var. *pallida.* Lighter and narrower.

5. V. ALPÉSTRIS (*inhabiting elevated situations*).

Sub-cylindrical, semi-transparent, glossy, light horn-colour, strongly striate in the line of growth; *whorls* four and a half, convex; *spire* short, blunt; *suture* excessively deep; *mouth* semi-oval and sub-angular; *teeth* four, visible from the outside from the thinness of the shell (one on the body, one on the columella, two within the outer lip); *umbilicus* small, but deep.

This shell differs from the last species in being more cylindrical, paler, and nearly transparent, more strongly striated, and it has no strengthening rib on the outer margin.

I may mention that I was the first to notice this

minute species in Ireland. In December, 1883, I found a single live specimen near Coleraine. As I was doubtful of its being actually *alpestris*, I sent it to Dr. Jeffreys, who kindly confirmed my opinion as to its identity. It is found, though rarely, in some of the northern counties of England.

6. V. SUBSTRIATÁ (*slightly striated*).

Ovate, thin, semi-transparent, glossy, very strongly and obliquely striate in the line of growth; *whorls* four and a half, cylindrical; *spire* very abrupt; *suture* remarkably deep; *mouth* semi-oval, outer margin constricted; *teeth* five to six (two or three on the body, one or two on the columella, two inside the outer lip) ; *umbilicus* small.

This pretty little shell is found in many parts of the British Isles in moist situations at the roots of grass and under stones. Its strong striations, almost amounting to ribs, serve to distinguish it from any of the previously mentioned species.

B. Sinistral.

7. V. PUSÍLLA (*minute*).

Fusiform, very glossy, thin, horn-colour; *whorls* four and a half; *spire* more tapering than that of most members of the genus; very slightly striate in the line of growth; *mouth* semi-oval, constricted in the outer margin, which is rather thick; *whorls* four and a half to five; *teeth* six to seven (two on the body, two on the columella, two or three inside the outer lip).

This species is found in many parts of England and Ireland, but it is local and rare. It frequents similar situations to the others.

8. V. ANGÚSTIOR (*narrower*).

Of the same form, but narrower than the last species, glossy, semi-transparent, horn-colour; closely striate in the line of growth; *whorls* four and a half, the penultimate being the broadest; *mouth* sub-triangular, and narrowed by the strong constriction of the outer margin; *teeth* four to five (two on the body, one on the columella, one inside the outer lip, which is thickened and reflected); *umbilicus* indistinct.

This species is very rare and local, occurring in a few localities only. In England at Bristol, Battersea Fields, and in Yorkshire; in Wales at Singleton (near Swansea), Tenby, Milford, and in some of the southern and western counties of Ireland. I have also found it abundantly on the sand-hills at the mouth of the River Bann.

The following table of comparison may be useful to assist identification :—

V. PUSILLA.	V. ANGUSTIOR.
Larger, broader in proportion.	Smaller, narrower in proportion.
Last whorl broadest.	Penultimate whorl broadest.
Mouth semi-oval.	Mouth triangular.
Outer lip very slightly contracted.	Outer lip very deeply contracted.
Teeth six to seven.	Teeth four to five.

C. DEXTRAL, *without teeth.*

9. V. EDÉNTULA (*toothless*).

Cylindrical, thin, glossy, horn-colour; slightly striated in the line of growth; *whorls* five and a half to six and a half; *spire* abrupt; *mouth* nearly circular; *lip* thin; *umbilicus* narrow.

Found in several parts of Great Britain and Ireland, but it is local.

It frequents decaying leaves and logs, and is particularly active after rain. Dr. Jeffreys remarks, and other writers bear him out in saying that "when crawling it usually carries its shell in a slanting position." It is, therefore, with much diffidence that I record my own observation as being opposed to this—in fact, what has struck me as noticeable is the perpendicular position of the shell when the animal is in motion; though my experience of the species is confined to a single locality in the north of Ireland, where it and its variety are abundant.

There is no mistaking this species for any other, as its shape and toothless mouth mark it off distinctly from any of the foregoing, and its size from the following species.

Var. *columella.* Longer, last whorl broader.

10. V. MINUTÍSSIMA (*very minute*).

Cylindrical, glossy, different shades of horn-colour; closely striated in the line of growth; *whorls* five and a half; *suture* deep; *mouth* similar to the last species, but somewhat quadrangular; *umbilicus* small.

This beautiful little shell is found in a few places in Scotland and England.

It is much smaller and narrower in proportion to its size than *V. edentula,* and is more strongly striated.

Genus VIII.—BÁLIA.

B. PERVÉRSA (*turned the wrong way*).

Sinistral, club-shaped, thin, dark horn-colour, semi-transparent, glossy, closely striated in the line of growth; *whorls* seven to eight; *mouth* squarish oval, sometimes furnished with a denticle on the columella; outer lip thin; *umbilicus* narrow.

This shell is local, though abundant where it occurs. It is found on the bark of trees; Dr. Jeffreys says "chiefly the beech, ash, sycamore, and apple." I once found it on a willow near Windsor, in company with *C. rugosa*, the young of which it much resembles. This resemblance it is necessary to draw attention to, or this local species may be passed over. It differs from the immature *C. rugosa* in being more slender, lighter in colour, having a deeper suture, and being without any carination along the basal ridge, which is a marked characteristic of the young of that species.

Var. *viridula*. Greenish white, transparent.

Genus IX.—CLAUSÍLIA.

This genus derives its name from a peculiar characteristic—a little door (Lat. *clausilium*).

This interesting provision of nature against the attacks of such enemies as beetles, &c., differs from an operculum thus:—It is not fastened to the *animal*, but to the pillar of the *shell* by an elastic filament. When the animal is within the shell this contractile filament draws the *clausilium* close over the aperture, about half a turn from the entrance, and when the animal emerges it pushes aside the little "spring-door," which then lies flat against the columella.

All the members of this genus are sinistral.

1. C. RUGÓSA (*wrinkled*).

Fusiform, purplish brown, but varying in colour; marked with small streaks of white; closely and irregularly striated in the line of growth; *whorls* ten to thirteen; *spire* tapering to a blunt point; *mouth* pear-shaped, expanding like a funnel, angulated above; *plications* two on the pillar, with two or three ribs between them, a plication behind the pillar lip, near which is a spiral fold; one or two teeth inside the outer lip; *umbilicus* narrow; *clausilium* shaped much like a "shoehorn."

This common shell is distributed throughout the whole of the British Isles, and is usually found on stone walls, rocks, rough bark of trees, and sometimes on sand-hills. It is very variable both in colour and markings. The smaller forms seem to be found, as might be expected, in northern and exposed situations. I have found it abundantly on the banks of the River Bann, adhering to stones which are completely covered at high tides.

Var. I. *albida*. Whitish.

Var. II. *Everetti*. Smaller.

Var. III. *gracilior*. More slender.

Var. IV. *tumidula*. Shorter, more ventricose.

Var. V. *dubia*. Larger, more ventricose.

Var. VI. *Schlechtii*. "Generally larger, more

elongated, smoother and more transparent than Var. *dubia*, pale brown, frequently resembling *C. laminata* in smoothness and transparency."

2. C. RÓLPHII (*after Rolph*).

Fusiform, thinner than the last species; reddish or yellowish brown, occasionally streaked with white; strongly striated; *whorls* nine to ten; the first two or three upper whorls are nearly of the same breadth, forming a short cylinder; *apex* blunt; *mouth* quadrangular; *plications* as in last species, but there are often two or three small teeth between the columellar folds; outer lip inflected; basal crest short and curved; *umbilicus* indistinct; *clausilium* oblong, regularly curved, slightly contracted above.

This is a rare and local species, found only in a few localities in some of the southern counties of England.

Dr. Jeffreys says: "From *C. rugosa* and its variety *dubia* this differs in being more ventricose and of a lighter colour, as well as in having much coarser striæ, which impart to the last-mentioned shell a decussated or slightly granular appearance. The mouth of the shell in *C. Rolphii* is, besides, larger and broader."

3. C. BIPLICÁTA (*having two folds*).

Fusiform, dull, but with a slight iridescence; rufous brown streaked with white; strongly striated; *whorls* twelve to thirteen; *mouth* oval, angular, contracted below, where a narrow but deep channel is formed; *teeth* as in *C. Rolphii; umbilicus* broader than usual in the genus; *clausilium* almost oval, slightly curved, attenuated below.

This is one of our very local shells. It is found at the roots and on the bark of willows by the Thames at Hammersmith not uncommonly; and also in Wilts.

It is much larger than *C. Rolphii*, and more slender in proportion. It is, moreover, streaked with white.

4. C. LAMINÁTA (*having plates*).

Similar in shape to the last species, pale reddish horn-colour; smooth and glossy; microscopically striated in the line of growth; nearly transparent; *whorls* twelve; *mouth* ovate; *teeth* very well marked on the columella; there are also three or four internal laminæ which are visible through the transparent shell; *umbilicus* small; *clausilium* shaped like a shoehorn, but oblong.

This handsome species is sparingly distributed throughout Great Britain as far north as Northumberland, and is also found in Ireland, but is local. It frequents the beech and ash, at the roots and on the trunks of which it may be found. It occurs in great abundance on the limestone rocks in Derbyshire.

Var. I. *pellucida.* Thinner, more transparent, highly glossy.

Var. II. *albida.* Greenish white.

5. C. PÁRVULA (*small*).

This shell differs from *C. rugosa* " in being smaller and quite smooth, with the exception of some very faint transverse lin'es, which are only observable with a lens, or of a few striæ near the mouth."

Specimens of this species, which is fairly common on the Continent, have been found near Stourbridge.

C. SÓLIDA (*solid*).

As only a single specimen of this Continental species has been found near some nursery gardens at Stapleton, near Bristol, whither it was doubtless brought with foreign plants, I think its claim to rank as British may be disregarded. Had it, like *T. Maugei,* " bred and multiplied," it might be viewed in a different light.

Genus X.—COCHLÍCOPA.

1. C. TRÍDENS (*with three teeth*).

Chrysalis-shaped, semi-transparent, solid, glossy, yellowish horn-colour, sometimes tinged with red or green; *whorls* seven; *spire* pointed; *mouth* narrow, with three denticles.

This shell is only found in England, where it is widely distributed, but local. It is gregarious, and frequents damp moss in shady spots.

Var. *crystallina.* Greenish white and transparent.

2. C. LÚBRICA (*slippery*).

Shaped much like the last species, but not so ventricose in proportion to its length; transparent, solid, greenish, sometimes with a reddish tinge; *whorls* five to five and a half; *spire* rounded at the point; *mouth* pear-shaped, and broader than the last species; *lip* often reddish.

This is a very common shell. It occurs in all parts of the British Isles in moss, under stones, logs, &c. It is a good plan to mount a row of these brilliant little objects on a strip of card. The varieties are numerous, as follows:—

Var. I. *hyalina.* Greenish white.

Var. II. *lubricoides.* Smaller and more slender.

Var. III. *viridula.* Shaped like Var. II., but greenish white.

Var. IV. *fusca.* Smaller, thinner, reddish brown.

Var. V. *ovata.* Much smaller, oval, spire shorter.

Genus XI.—ACHATÍNA.

A. ACÍCULA (*a hair-pin*).

Slender and tapering in shape, semi-transparent, white, thin, and very glossy; *whorls* five and a half; *spire* obtuse; *apex* rounded; *suture* distinct; *mouth* long and rounded at the base, which has a deep notch.

This interesting animal is the only British member of a genus consisting of nearly one hundred and sixty species. It lives underground, and is never found on the surface of the earth in a live state. Whether its subterranean habits are the cause or the effect of its being destitute of the power of sight need not be discussed here, but it is the fact that, in common with other subterranean animals, it is eyeless.

Dr. Jeffreys remarks: "In all probability the *A. acicula* lives upon animal matter; for, in the spots where it has been found living, no underground fungus or other vegetation appears to exist, and the form of the shell would induce a belief that this snail is not only zoophagous but predaceous. The shells of all true *Glandina*, which are carnivorous, have the same kind of notch or truncature at the base as the present species of *Achatina*."

It inhabits various parts of England, Wales, and Ireland, under stones and at the roots of grass, some inches underground. As it would be rather a tedious process to search for it by digging all over a district, it is perhaps fortunate that another means of obtaining it exists. Among the rejectamenta of rivers (Thames, Yorkshire Ouse, &c.) specimens are common, being doubtless washed away by floods from their native localities. The specimens found in this way are of course dead, but this shell fortunately does not suffer much in appearance from exposure.

It must be borne in mind that specimens found thus on the river bank do not indicate the actual locality of the shell, for who shall say how far they have travelled before being stranded?

Family IV.—CARYCHIIDÆ.

Genus.—CARÝCHIUM.

C. MÍNIMUM (*smallest*).

Sub-fusiform, semi-transparent, white, rather solid, glossy; microscopically striate in the line of growth; *whorls* five and a half; *spire* pointed; *mouth* ovate, furnished with three denticles; *umbilicus* small and narrow.

There is no mistaking this little shell when once seen. It must, however, be examined under a lens in order to be fully appreciated. It is worth while to wait for the animal to come out and travel over a moist decayed leaf, when the conspicuous black specks that constitute its eyes can be seen and their sessile position noted; indeed, these eyes may be seen *through* the shell, so dark are they and so transparent the shell.

It is very common under stones, logs, &c., at the roots of grass, and particularly among moist decaying leaves in all parts of the British Isles south of the Moráy Firth.

Family V.—CYCLOSTOMATIDÆ.

Genus I.—CYCLÓSTOMA.

C. ÉLEGANS (*elegant*).

Globosely oval, solid, opaque, light pink, marked with streaks or spots of a darker colour; very strongly striated in the line of growth, and more closely striated spirally; *whorls* four and a half, cylindrical; *suture* very deep; *spire* blunt; *mouth* nearly circular; *umbilicus* small, but deep; *operculum* thick, testaceous.

This interesting shell is the only British representative of an interesting genus which abounds in species in warmer latitudes.

It is common in England, Wales, and Ireland, mostly—though not exclusively—in calcareous soils near the sea. Along the banks of the Thames between Marlow and Maidenhead it occurs in great abundance on the chalk hills. It is also tolerably plentiful in Jersey, where there are no calcareous strata.

Genus II.—Ácme.

A. LINEÁTA (*lined*).

Cylindrical, light brown, slightly transparent, rather glossy, deeply striate in the line of growth; *whorls* six to seven; *spire* blunt; *suture* distinct; *mouth* pear-shaped; *umbilicus* very small; *operculum* horny.

This, the most minute of operculated land shells, though widely distributed throughout all the British Isles, is local. It may be found in woods, under stones, and on decaying leaves.

It is said to be gregarious.

Var. *alba*. White or colourless and transparent.

APPENDIX A.

SUB-KINGDOM.	CLASS.	ORDER.	FAMILY.	GENUS.
Mollusca.	1. Conchifera (1) (*bivalves*).	1. Lamellibranchiata (1).	1. Sphæridæ (1).	1. Sphærium (1).
				2. Pisidium (2).
			2. Unionidæ (2).	1. Unio (3).
				2. Anodonta (4).
			3. Dreissenidæ (3).	1. Dreissena (5).
	2. Gasteropoda (2) (*univalves*).	1. Pectinibranchiata (2).	1. Neritidæ (4).	1. Neritina (6).
			2. Paludinidæ (5).	1. Paludina (7).
				2. Bythinia (8).
				3. Hydrobia (9).
			3. Valvatidæ (6).	1. Valvata (10).
		2. Pulmonobranchiata (3).	1. Limnæidæ (7).	1. Planorbis (11).
				2. Physa (12).
				3. Limnæa (13).
				4. Ancylus (14).
			2. Limacidæ (8).	1. Arion (15).
				2. Geomalcus (16).
				3. Limax (17).
			3. Testacellidæ (9).	1. Testacella (18).
			4. Helicidæ (10).	1. Succinea (19).
				2. Vitrina (20).
				3. Zonites (21).
				4. Helix (22).
				5. Bulimus (23).
				6. Pupa (24).
				7. Vertigo (25).
				8. Balia (26).
				9. Clausilia (27).
				10. Cochlicopa (28).
				11. Achatina (29).
			5. Carychiidæ (11).	1. Carychium (30).
			6. Cyclostomatidæ (12).	1. Cyclostoma (31).
				2. Acme (32).

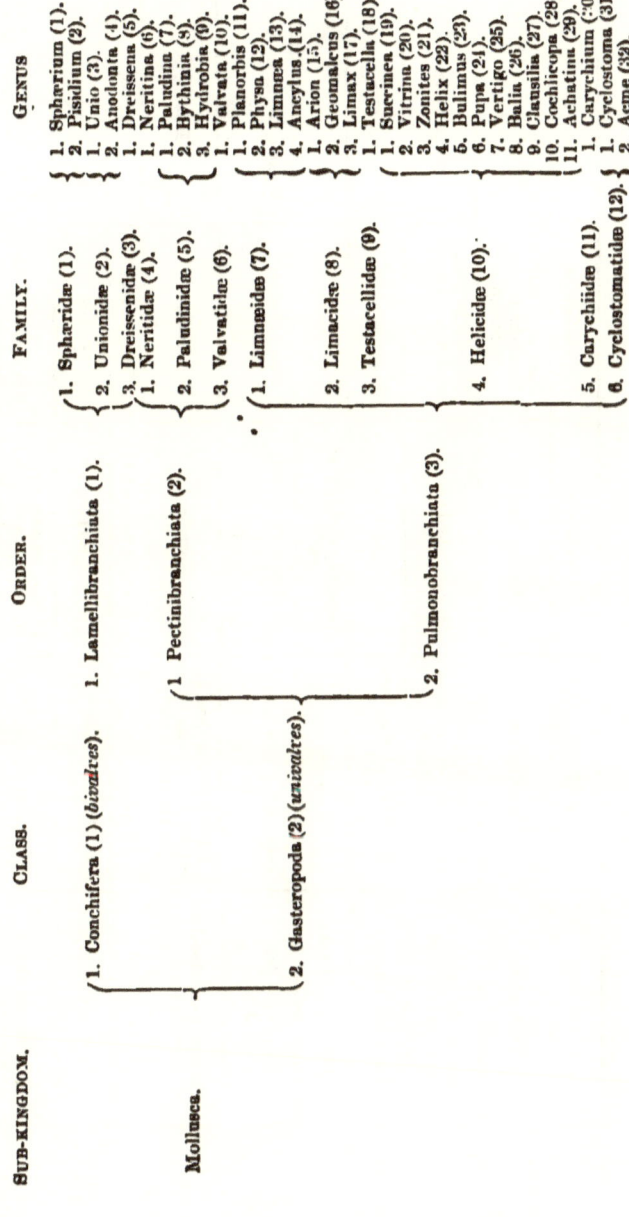

Allied Zonites.

	1. *cellarius.*	2. *glaber.*	3. *alliarius.*	4. *nitidulus.*	5. *nitidus.*
ANIMAL. Body.	Slate grey.	Bluish grey.	Darker than 1.	Dark grey with a brownish tinge, speckled.	Slaty black, coarsely speckled.
Foot.	Yellowish, pointed, narrow, keeled behind.	Sole separated from upper part of foot by a dark line.	Very long and narrow.	Narrow in front, swollen and keeled behind.	Narrow, keeled.
Tentacles.	Long and slender.	Upper long, lower very short.	Short.	Short, bulbs small.	Thick.
Smell.	Said to smell of garlic.	—	Smelling strongly of garlic.	—	—
SHELL. Whorls.	Spire very flat, 5-6.	Spire raised to a point, 5-5½.	Spire more raised than 1. 4-5.	Spire slightly convex. 4-5.	Spire convex, suture deep. 5.
Mouth.	Obliquely and deeply semi-lunar.	More circular than 1, but more interrupted by penultimate whorl.	Narrow.	Round.	Round.
Umbilicus.	Rather broad.	Less open than 1.	More open than 1.	Very broad.	Narrow.
Diameter of circumference.	Nearly ½ inch.	Slightly more than ⅜ inch.	¼ inch.	¼ inch.	¼ inch.
Colour and texture.	Pale brown above, opaque white below.	Clearer than 1, less white below.	Thicker in proportion, and more glossy than 1; darker, with less white below.	Not so glossy, nor so transparent as 1.	Clear chocolate brown, very glossy, not white below; strongly striate.
Most distinctive characteristic.	Largest species.	Spire raised to a point, mantle visible through the shell.	Smell of garlic.	Umbilicus very large.	No white below, strong striations, animal nearly black.

APPENDIX C.

Vertigo.—Dextral, denticulated.

	1. *antivertigo.*	2. *Lilljeborgii.*	3. *Moulinsiana.*	4. *pygmæa.*	5. *alpestris.*	6. *substriata.*
General appearance and texture.	Dark reddish, very glossy.	Same as 3.	Very pale, very glossy, nearly transparent.	Fairly glossy.	Very glossy, thin.	Fairly glossy.
Striations.	Faint and close in line of growth.	Same as 3.	Microscopical, in line of growth.	Faint, in line of growth.	Rather strong in line of growth.	Very strong, oblique, in line of growth.
Teeth.	*Reddish in colour.* 6-9. { 2 or 3 on body, 1 or 2 on columella, 3 or 4 inside lip.	4. { 1 on body, 1 on columella, 2 inside lip.	4-5. { 1 on body, 1 on columella, 2 or 3 inside lip.	Same as 3.	4. { 1 on body, 1 on columella, 2 inside lip.	5-6. { 2 on body, 1 or 2 on columella, 2 inside lip.
Locality and distribution.	Throughout the Brit. Isles S. of Moray Firth.	West of Ireland. Rare and local.	Hertfordshire and Hampshire. Rare and local.	Brit. Isles generally. Common.	Northern counties of Eng. and N. Ireland. Rare.	Gt. Britain from Skye to Devon, and throughout Ireland.
Most distinctive characteristics.	Large size, teeth reddish.	More swollen than 3, labial rib stouter, often one tooth less.	Much the largest of all except 2. Tumid, pale, transparent.	Usually 5 teeth, oval form.	Lip thin, unreflected; apex blunt; suture deep; striations strong; form cylindrical.	2 teeth on body, strong oblique striations.

GLOSSARY

OF TECHNICAL TERMS ACCENTUATED.

A.

acéphalous, headless, applied to bivalves.
achatína, agate.
acícula, a hair-pin.
ácme, a point.
aculeáta, prickly.
acúminata, pointed.
acútus, pointed.
agréstis, inhabiting fields.
álbida, white.
albína, white.
albo-fasciáta, having white bands.
álbus, white.
alliárius, smelling of garlic.
alpéstris, inhabiting heights.
ámnicum, inhabiting rivers.
amphíbious, inhabiting both land and water.
ampulláceous, shaped like a flask or bottle.
anatína, belonging to ducks.
áncylus, hooked.
angústior, narrower.
anodónta, toothless.

antérior, the front; applied to bivalves, the large or blunt end.

antivertigo, not reversed.

áperture, an opening ; of univalves the mouth.

ápex, the extreme point of the spire of a univalve.

aquátic, inhabiting water.

árborum, inhabiting trees.

arbustórum, inhabiting copses.

Aríon. A fabled musician. The name of a genus.

aspérsa, spotted, sprinkled.

áter, black.

atro-purpúrea, dark purple.

auriculúria, ear-shaped.

áxis, applied to univalve shells, the centre column formed by the junction of the whorls.

B.

Bália, striped.

beak, see *umbo*.

bidentáta, having two teeth.

bigranáta, having two small teeth.

bilóbed, divided into two rounded parts.

biplicáta, having two folds.

bivalve, applied to shells with two movable plates forming the sides.

bizóna, with two bands.

bránchial, pertaining to the gills.

Búlimus (deriv. disputed), the name of a genus.

Burnétti, after Burnet.

býssus, strands by which *D. polymorpha*, &c., attach themselves to stones.

Bythínia. An ancient province of Asia Minor ; the name of a genus.

C.

calcáreous, chalky.

Cantiána, inhabiting Kent.

caperáta, wrinkled.

capuloídes, resembling a handle.

cárdinal, applied to bivalves, belonging to the hinge.

carinátus, keeled.

carnivorous, eating animal matter.

Cartusiána, named after a Carthusian monastery.

Carýchiídœ. A Family of mollusks, from the typical genus *Carychium*.

Carýchium. A genus named after a marine shell.

cellárius, inhabiting cellars.

cinérea, ash coloured.

Clausília. The name of a genus with an appendage called a *clausilium*. (See page 87.)

Cochlícopa, a shell with a notch.

columélla, a small column; the technical term for the *axis* (q. v.).

compácta, compact.

complanátus, flattened.

compréssa, compressed.

cóncave, hollow.

concéntric, having the same centre.

Conchífera. Shell-bearing; the name of the Class comprising bivalves.

Conchólogy. "Conchology treats of the Mollusca, or that great division of invertebrate animals which have soft bodies and an organization superior to that of insects and only inferior to that of fishes."

concínna, neat.

cónica, conical.

conoídea, cone-like.

contécta, covered.

contórtus, twisted together.

contráctile, able to become shorter.

cónvex, bulging out.

córneus, horny.

córvus, a crow, applied to a very dark var. of *L. palustris.*

costáta, ribbed.

cristáta, crested.

crystállinus, crystalline.

cúrta, short.

curviróstris, with curved beaks.

Cyclóstoma, round mouth ; name of a genus.

Cyclóstomatidæ. Name of a Family of mollusks, from the typical genus *Cyclostoma.*

cygnea, belonging to swans.

D.

decíduous, liable to fall off.

decolláta, shortened from the top, truncated.

decússated, intersected by cross-lines.

dénticle, a small tooth.

dentition, see *lingual.*

depréssa, flattened.

depressiúscula, somewhat flattened.

déxtral. A univalve is said to be dextral when the mouth opens to the observer's right as he holds it with the spire pointing upwards.

dextrórsa, turned to the right hand.

dilatátus, expanded.

Dillwýnii, named after Dillwyn.

discifórmis, formed like a circular plate.

dórsal, belonging to the back.

Drápanáldi, named after Drapanaud.

Dreisséna, named after Dreissens.

dúbia, doubtful.

E.

edéntula, toothless.

élegans, elegant.

elongáta, lengthened.

epidérmis, the thin skin covering shells.

épiphragm, a film which is secreted by the animal to cover the mouth of the shell.

equiláteral, having equal sides.

ericetórum, inhabiting heaths.

Everétti, named after Everet.

exálbida, whitish.

excavátus, hollowed out.

F.

fílament, a thread.

flavéscens, yellowish.

flávus, yellow.

fluviátilis, inhabiting rivers.

fontinále, inhabiting springs.

foot, a flexible muscular process by means of which mollusks travel.

frágilis, frail.

fúlvus, tawny.

fúsca, dark brown.

fúsiform, spindle-shaped.

G.

gagátes, jet.

Gasterópoda. A Class of univalve mollusks, the lower surface of whose belly forms the foot.

génus. In natural history a subdivision of an *Order.* A genus is subdivided into *species.* (See Appendix A.)

Geomálcus, an earth mollusk; the name of a genus.

gibbósa, swollen.

gíbbous, protuberant.

gláber, smooth.

glutinósa, sticky, slimy.

gracílior, rather slender.

gránulated, with small markings like grains.

gregárious, living in flocks.

H.

hábitat, dwelling place.

háliotídea, ear-shaped; from *Haliotis,* or "Venus' ear."

Helicídæ. A Family of mollusks.

Hélix, a screw, coil; the name of a genus.

Hélmii, named after Helm.

Henslowána, named after Prof. Henslow.

herbívorous, feeding on vegetable matter.

híspid, hairy.

horténsis, inhabiting gardens.

hýaline, glassy.

hybrida, a cross between two species.

hypnórum, living among the moss *Hypnum.*

I.

inconspícua, not easily observed.

incrassáta, thickened, coarse.

indígenous, born in the country, native.

infláta, blown out, swollen.

instábilis, fluctuating.

intermédia, varying between two distinct forms.

involúta, folded inwards.

iridéscent, coloured like a rainbow.

L.

labiósa, with a large lip.

lacústris, inhabiting lakes or ponds.

lévis, smooth.

lamelláta, furnished with plaits.[1]

Lamellibránchiáta, having leaf-like gills; the name of an Order of mollusks.

lámina, a plate.

lamináta, furnished with plaits.

lapicída, a stone-cutter.

látior, broader.

Léachii, named after Dr. Leach.

leúcozóna, banded with white.

lígament, the elastic substance connecting the valves of a bivalve shell.

Límax, a slug; the name of a genus.

Limacidæ. The name of a Family of mollusks.

Límnæa, inhabiting marshes; the name of a genus.

Limnæidæ. The name of a Family of mollusks.

lineátus, marked with lines.

lingual dentition. The arrangement of teeth on the tongue. The gasteropods are furnished with a "ribbon" set with teeth, which grind against a hard palate.

---

[1] *Lamina* means a *plate*; *lamellata* and *laminata* are used erroneously to describe *plaits*, *plications* or *folds*.

lip. The lip of a shell is the outer edge of the mouth.

lubricoídes, somewhat smooth.

lútea, yellowish.

M.

maculósus, spotted.

májor, larger.

mantle, a flexible outgrowth of the body, resembling a hood, which contains glands that secrete the colouring matter of the shell.

margaritácea, pearly.

marginátus, having a rim or border.

marítima, living on the sea coast.

Maúgei, named after Mauge.

máximus, largest.

membranáceous, composed of fine-spun fibres.

micróstoma, having a small mouth.

mínimus, smallest.

mínor, smaller.

minutíssima, very minute.

mollúsk, } a soft-bodied animal.   (See *Conchology.*)
mollúsc, }

montánus, inhabiting mountains.

Mortoni, named after Morton.

Moulinsiána, named after M. des Moulins.

múcus, slime.

múcronáta, pointed.

N.

nácreous, pearly.

nána, dwarf-like.

Nautiléus, shaped like a nautilus.

Nelsóni, named after Nelson.

nemorális, inhabiting groves.

Neritidæ. A Family of mollusks.

Neritína. The name of a genus.

nítens, shining.

nitídulus, rather shiny.

nítidus, shiny.

núcleus, the tip of the spire of a univalve shell.

O.

oblónga, oblong.

obscúrus, hidden.

obtusális, blunt.

obvolúta, wrapped up.

ochrácea, yellow.

opérculated, furnished with an *operculum.*

opérculum, a covering or lid. A hardened plate attached to the foot of the animals of many univalves, which closes like a door or lid when the animal withdraws into the shell.

Order. In natural history the subdivision of a Class.

órifice, a hole.

ornáta, ornamented.

ovális, oval.

ovíparous, producing eggs.

óvovivíparous, producing eggs which are hatched internally.

P.

pállida, pale.

Paludína, inhabiting marshes; the name of a genus.

Paludinidæ. A Family of mollusks.

palústris, living in marshes.

Pectinibránchiáta, having gills resembling a comb; the name of an Order of mollusks.

pédicle, a foot-stalk.

pellúcida, transparent.

penúltimate, the last but one.

péregra, a traveller.

períphery, applied to univalve shells, the outline of the body-whorl.

péristome, the rim of the mouth of univalve shells.

pervérsa, turned the wrong way.

Phýsa, a bubble; the name of a genus.

picta, painted, ornamented.

pictórum, belonging to painters.

Pisána, named after the town of Pisa.

pisidióides, somewhat resembling a *Pisidium*.

Pisidium, resembling a pea; the name of a genus.

Planórbis, a flat coil; the name of a genus.

plicátion, a fold; used for the thickened process which resembles a tooth in the mouth of a univalve.

pomátia, having a lid or covering.

polymórpha, many-shaped.

postérior, applied to bivalve shells, the thinner or sharper end.

pulchélla, minutely beautiful.

Pulmonobránchiáta, having gills resembling lungs; an Order of mollusks.

Púpa, a doll or chrysalis; the name of a genus.

púrus, clear.

pusillum, small.

pútris, stinking.

pygmǽa, minute.

pyramidális, shaped like a pyramid.

pýriform, pear-shaped.

R.

radiáta, rayed.

radiátulus, slightly rayed.

reflécted, bent back.

retículated, like network.

reveláta, unveiled, discovered.

ríngens, grinning like a dog, *i.e.* showing the teeth.

rivícola, an inhabitant of streams.

róseo-labiáta, having a pink-coloured lip.

róseum, pink.

rostráta, beaked.

rotúnda, round.

rotundáta, rounded.

ruféscens, reddish.

rufilábris, having red lips.

rugósa, wrinkled.

rupéstris, inhabiting cliffs.

S.

scálarifórmis, formed like a *Scalaria* (a spiral sea shell with projecting ribs which give the appearance of the divisions of a ladder, *scala*).

scar. "Muscular scars" are the depressions formed by the attachment of the muscles holding the parts of a bivalve shell together.

scútulum, a small shield.

secále, a grain of rye.

séptum (plural *sépta*), a division.

serícea, silky.

séssile, situated on a flat surface, not raised on a stalk.

shagreéned, covered with small granules.

símilis, resembling (another species).

sinístral. The opposite to *dextral* (q. v.).

sinistrúrsa, turned to the left.

sinuáta, curved.

síphon, a tube or pipe.

sólida, solid.

solídula, somewhat solid.

spécies, the subdivision of a genus.

specífic, belonging to a species.

Sphærídæ. A Family of mollusks.

Sphærium, shaped like a globe ; the name of a genus.

spire, applied to a univalve shell, all the whorls except the lowest.

spirórbis, having a circular spire.

spléndens, shining.

stagnális, inhabiting marshes.

stríæ, fine lines.

striated, having fine lines.

sub-apérta, somewhat open.

sub-cylíndrica, somewhat cylindrical.

sub-globósa, somewhat rounded.

sub-marítima, living almost on the sea coast.

sub-rúfa, somewhat red.

sub-scaláris, somewhat *scalariform* (q. v.).

sub-striáta, somewhat *striated* (q. v.).

Succinea, amber-coloured ; name of a genus.

succineæförmis, shaped like a *Succinea.*

súture, applied to univalves, the division between the whorls.

sýnonym, a name that has the same meaning as another name.

T.

tenéllus, tender.

tentaculáta, furnished with tentacles.

ténuis, thin, slender.

terréstrial, inhabiting the land.

Testacélla, a small shell; the name of a genus.

Testacellidæ. A Family of mollusks.

testáceous, having a hard shell.

tíncta, dyed.

trídens, furnished with three teeth.

trúncate, cut short, ending abruptly.

truncátula, slightly truncate.

tubérculate, furnished with pimples.

túmidus, swollen.

Túrtoni, named after Dr. Turton.

týpe, a standard.

týpical, resembling the type.

U.

umbilicáta, having an umbilicus.

umbilícus, the navel; applied to univalve shells, the cavity formed by the whorls when they do not form a solid axis or columella.

úmbo, the knob in the centre of a shield; applied to bivalve shells, the umbones (or beaks) are the protuberances by the hinge which constituted the infant shell.

unícolor, of one colour.

Únio, a pearl; the name of a genus.

Unionidæ. A Family of mollusks.

únivalve, (a shell) consisting of one piece.

V.

Valváta, having a valve (*i.e.* operculum); the name of a genus.

Valvatídæ. A Family of mollusks.

válve, a complete part of a shell.

variety. Varieties are members of a species that deviate constantly from the type in form, size, or colour.

ventricósa, full-bellied, swollen.

ventrósa, swollen.

Vertígo, a twisting; the name of a genus.

viréscens, greenish.

virgáta, striped.

viridescénti-álba, greenish white.

viridula, somewhat green.

vítrea, glassy.

vítrina, transparent like glass.

vivípara, producing young alive.

vórtex, a whirlpool.

W.

whorl, a twist of a spiral shell.

Z.

Zellénsis, after a town named Zell.

Zonítes, circular, like a girdle; the name of a genus.

INDEX.

The first numbers indicate the page, the second the plate, and the third the figure.

REFERENCE TO PLATES.

I.

1. Sphærium corneum.
2.   ,,   rivicola.
3.   ,,   ovale.
4.   ,,   lacustre.
5. Pisidium amnicum.
6.   ,,   fontinale.

7. Pisidium pusillum.
8.   ,,   nitidum.
9.   ,,   roseum.
10. Unio tumidus.
11.   ,,   pictorum.
12.   ,,   margaritifer.

II.

1. Anodonta cygnea.
2.   ,,   anatina.
3. Dreissena polymorpha.
4. Neritina fluviatilis.
5. Paludina contecta.
6.   ,,   vivipara.

7. Bythinia tentaculata.
8.   ,,   Leachii.
9. Valvata piscinalis.
10.   ,,   cristata.
11. Hydrobia similis.
12.   ,,   ventrosa.

III.

1. Planorbis lineatus.
2.   ,,   nitidus.
3.   ,,   Nautileus.
4.   ,,   albus.
5.   ,,   glaber.
6.   ,,   spirorbis.
7.   ,,   vortex.

8. Planorbis carinatus.
9.   ,,   complanatus.
10.   ,,   corneus.
11.   ,,   contortus.
12.   ,,   dilatatus.
13. Physa hypnorum.
14.   ,,   fontinale.

IV.

1. Limnœa glutinosa.
2.   „   involuta.
3.   „   peregra.
4.   „   auricularia.
5.   „   stagnalis.
6.   „   palustris.
7.   „   truncatula.
8.   „   glabra.
9. Ancylus fluviatilis.
10.   „   lacustris.

11. Limax gagates.
12.   „   marginatus.
13.   „   flavus.
14.   „   agrestis.
15.   „   levis.
16.   „   arborum.
17.   „   maximus.
18. Testacella Haliotidea.
19.   „   Maugei.

V.

1. Succinea putris.
2.   „   elegans.
3.   „   oblonga.
4. Vitrina pellucida.
5. Zonites cellarius.
6.   „   alliarius.
7.   „   glaber.
8.   „   nitidulus.
9.   „   purus.

10. Zonites radiatulus.
11.   „   nitidus.
12.   „   excavatus.
13.   „   crystallinus.
14.   „   fulvus.
15. Helix lamellata.
16.   „   aculeata.
17.   „   memoralis.
18.   „   arbustorum.

VI.

1. Helix pomatia.
2.   „   aspersa.
3.   „   Cantiana.
4.   „   Cartusiana.
5.   „   rufescens.
6.   „   concinna.
7.   „   hispida.
8.   „   sericea.

9. Helix revelata.
10.   „   fusca.
11.   „   Pisana.
12.   „   virgata.
13.   „   caperata.
14.   „   ericetorum.
15.   „   rotundata.

VII.

1. Helix rupestris.
2. „ pygmæa.
3. „ pulchella.
4. „ lapicida.
5. „ obvoluta.
6. „ villosa.
7. Bulimus acutus.
8. „ montanus.
9. „ obscurus.
10. Pupa secale.
11. „ ringens.

12. Pupa umbilicata.
13. „ marginata.
14. Vertigo antivertigo.
15. „ Moulinsiana.
16. „ pygmæa.
17. „ alpestris.
18. „ substriata.
19. „ pusilla.
20. „ angustior.
21. „ edentula.
22. „ minutissima.

VIII.

1. Clausilia rugosa.
2. „ Rolphii.
3. „ laminata.
4. „ biplicata.
5. Balia perversa.
6. Cochlicopa lubrica.

7. Cochlicopa tridens.
8. Achatina acicula.
9. Carychium minimum.
10. Cyclostoma elegans.
11. Acme lineata.

IX.

Plate explaining the technical names for the different parts of a bivalve and a univalve shell.

CHISWICK PRESS :—C. WHITTINGHAM AND CO., TOOKS COURT, CHANCERY LANE.

Plate 1.

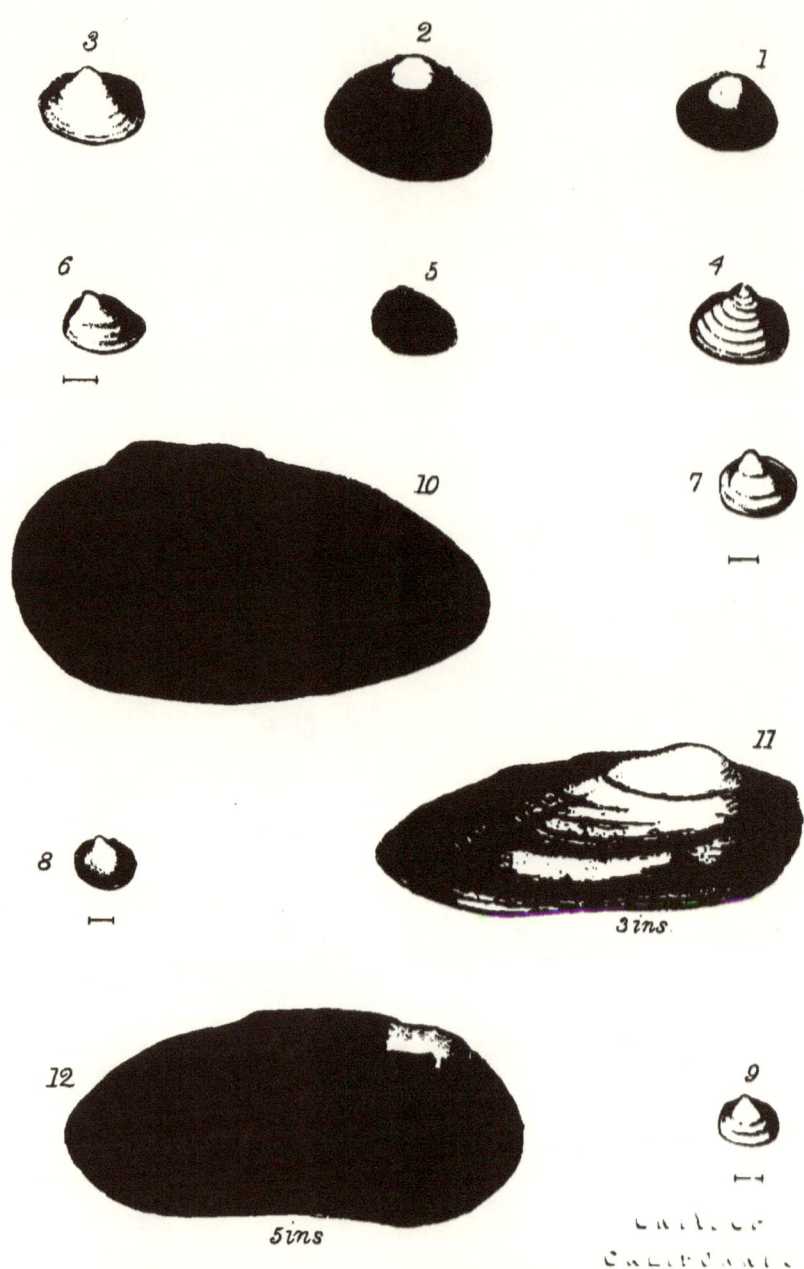

G.W. Adams del.

G. Jarman sc

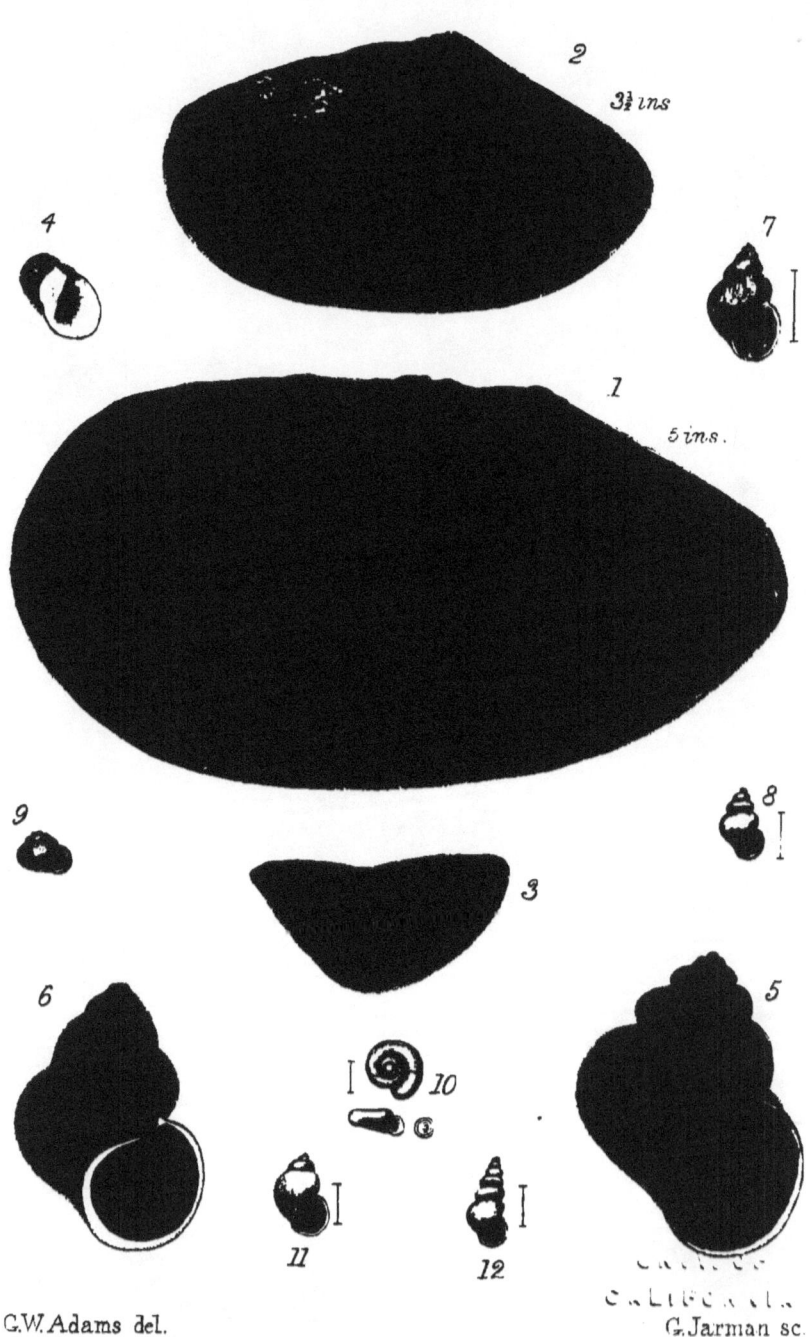

Plate II.

G.W. Adams del.

G. Jarman sc.

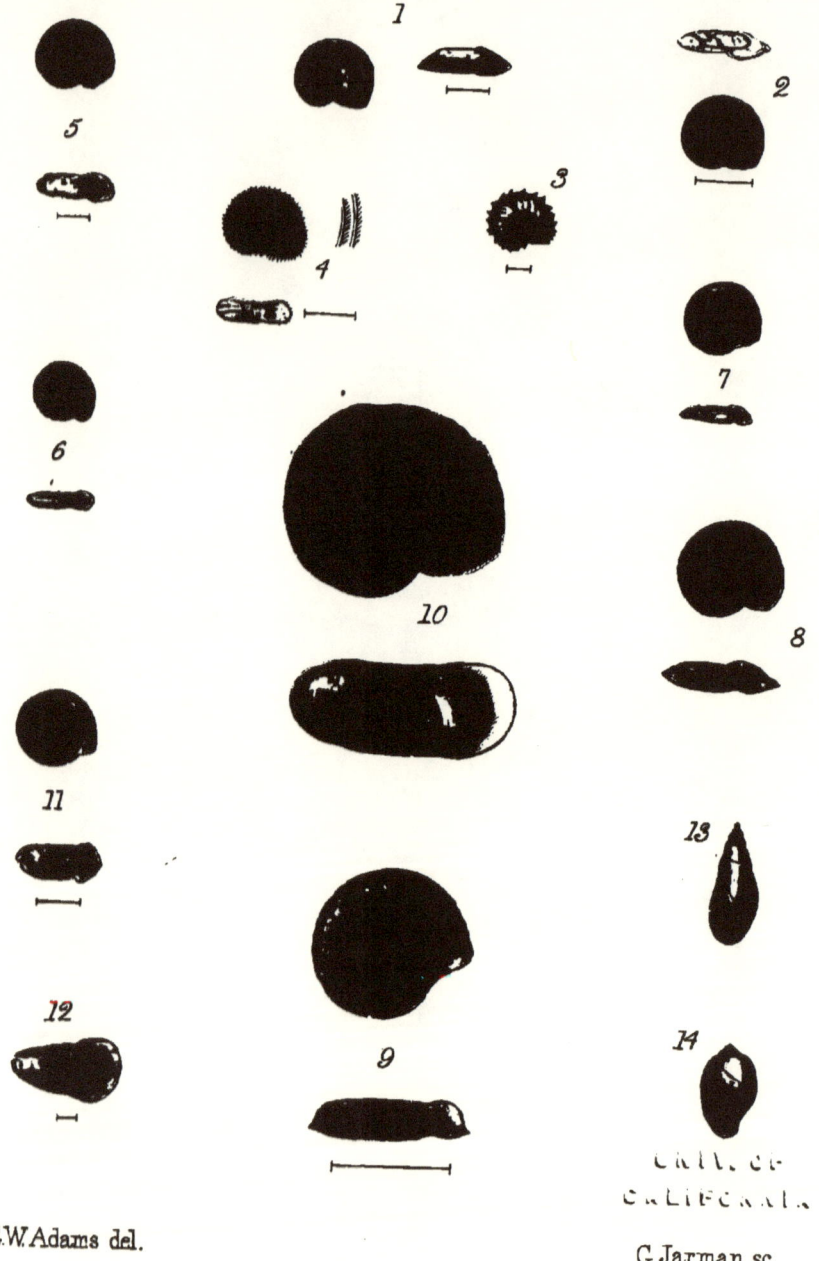

G.W. Adams del.

G. Jarman sc.

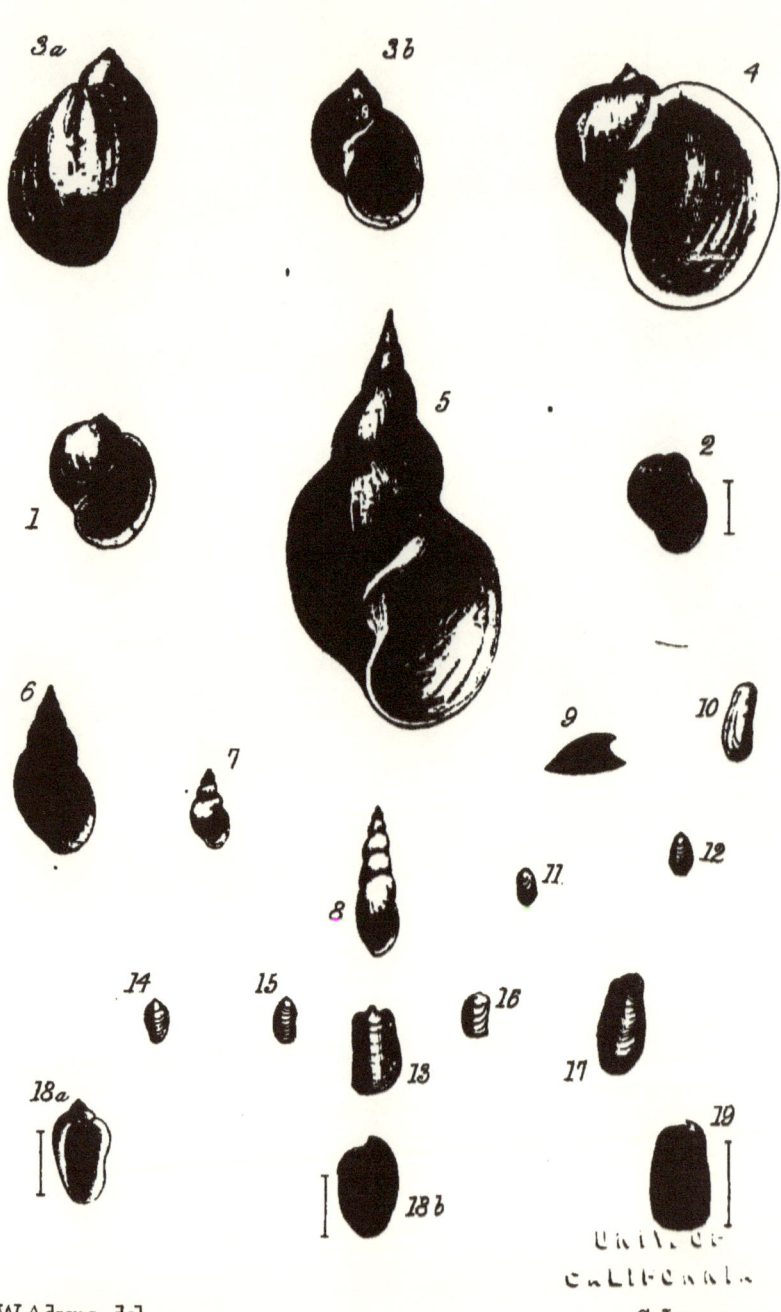

Plate IV

G.W. Adams del.

C. Jarman sc.

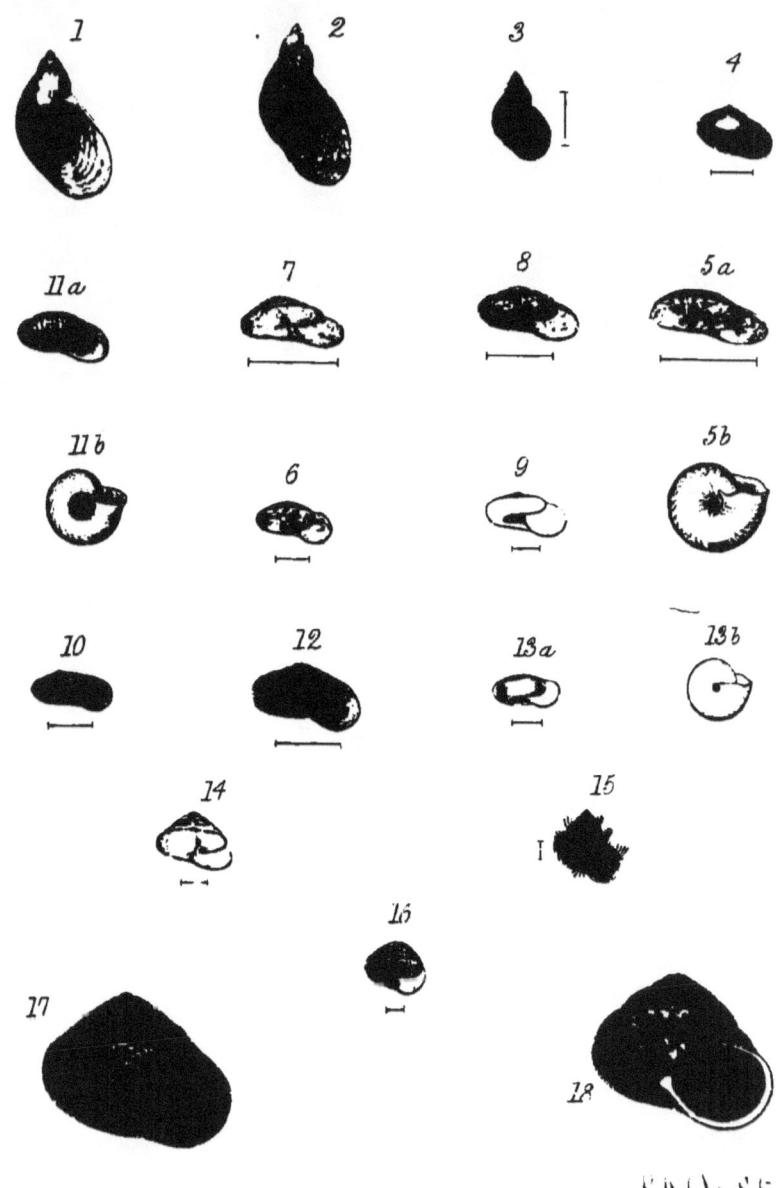

G.W.Adams del.

G.Jarman sc.

Plate VI.

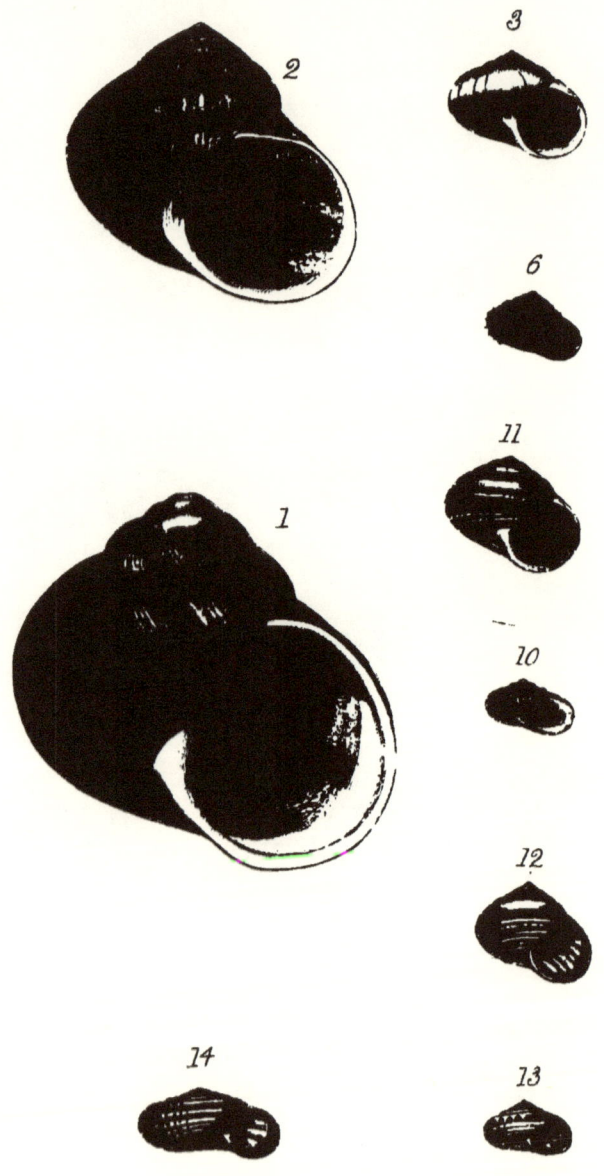

G.Jarman sc.

Plate VII.

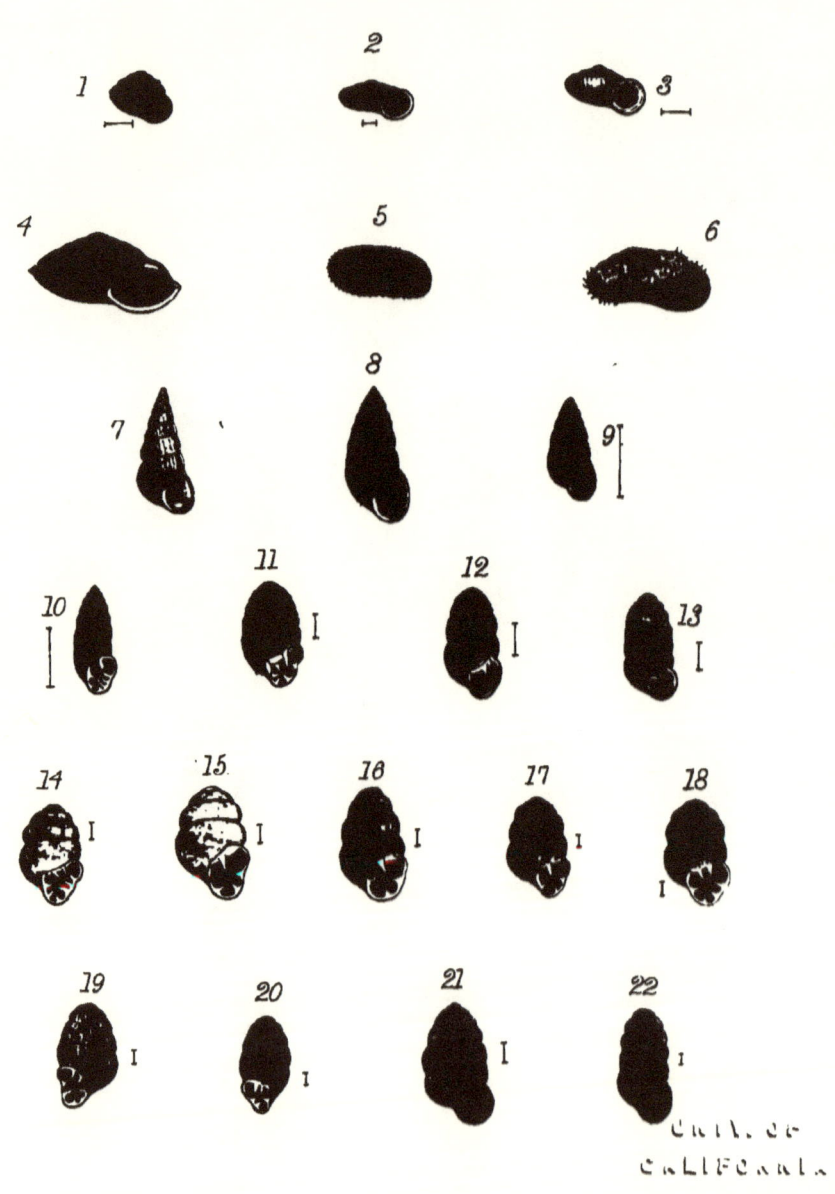

C.W.Adams } del.
L.F.Adams }

G.Jarman sc.

Plate VIII

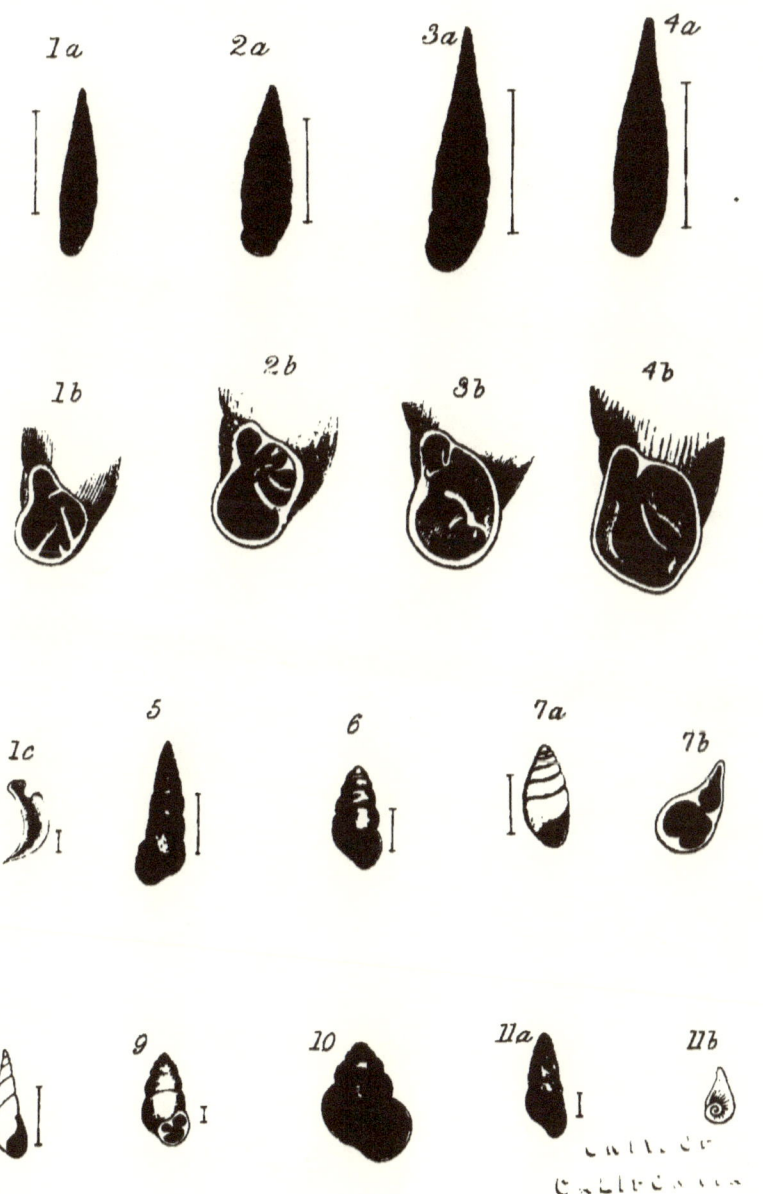

G.W.Adams del.   G.Jarman sc.

www.ingramcontent.com/pod-product-compliance
Lightning Source LLC
Chambersburg PA
CBHW030901050726
47500CB00009B/557